North Stradbroke Island

For Gladys Rahnsleben, whose early records have been invaluable in preparing this work.

North Stradbroke Island

Ellie Durbidge, Jeanette Covacevich

2nd Edition (revised 2004)

Stradbroke Island Management Organisation
North Stradbroke Island, Queensland

First edition 1981
Second edition 2004

Printer:
Watson Ferguson & Company
Moorooka, Qld 4105

National Library of Australia
Cataloguing in publication data
Durbidge, Ellie
North Stradbroke Island

ISBN 0 9593983 0 9

I Natural resources – Queensland – North Stradbroke Island
II North Stradbroke Island (Queensland) – Description and travel
III Covacevich, Jeannette II Stradbroke Island Management Organisation III Title

333.78'09943'2

Illustrations: Ellie Durbidge
Photographs: Bruce Cowell
Design of first edition: Jane Van Dyck
Cover design: Jane Van Dyck
Cover photograph: Bruce Cowell
Dedication page photograph (sand dunes, Main Beach): the late Bob Pattison

CONTENTS

ACKNOWLEDGMENTS

We have had a lot of help in preparing this work. Information was supplied or checked by Doug Anning, Howard Bell, the late Grandfather Borey, Captain H G Chesterman, the late Miss Mabel Brown, Francis Clayton, Mrs B Crouchley, Greg Czechura, the late Bonty Dickson, Tony Durbidge, Selwyn Everist, Jeff Ford, Wendy Goebel, Alan Gunn, Roger Hardley, Glen Ingram, the late Mr Willie McKenzie, Joe Mendez, Geoff Monteith, Miss M O'Keefe, Eve Pattison, Mrs Vera Perry, Gladys and Norman Rahnsleben, Ron Stubbin, Miss Anastasia Tripcony, Mr Stephen Van Dyck, and Kath Walker. Officers of the Oxley and Mitchell Libraries, the Queensland State Archives, the Department of Mapping and Survey, the Department of Aboriginal and Islanders Advancement, National Parks and Wildlife Service, the Queensland Museum, the Redland Shire Council, and the Queensland Fisheries Service have also assisted us by providing or checking data.

Norman Rahnsleben compiled the Rainfall Appendix (33 years rainfall records for Point Lookout).

The poem 'Minjerriba' is included with the permission of its author, the late Oogeroo Noonukul/Kath Walker.

Bruce Campbell and Richard Robins of the Queensland Museum read the manuscript or parts of it critically. Their many suggested modifications have improved the text.

Typing was done by Debbie Vale, Miss Joyce Stewart, and Dianne Ashwell.

Jeanette Covacevich thanks the Director and the Board of Trustees of the Queensland Museum for permission to be involved in this project.

1981

ACKNOWLEDGMENTS (REVISED EDITION)

We thank the following for their assistance:

Frances Quinn, Stephen Van Dyck, Harry Hines, Mike Hines, Greg Litherland, Jane Van Dyck, Redland Shire Council, Natural Resources and Mines, Queensland Fisheries Services, Consolidate Rutile Limited, North Stradbroke Island Historical Museum, Queensland Herbarium, Queensland Museum, Queensland National Parks and Wildlife Services.

Without the stalwart efforts of Haig Beck and Jackie Cooper, who undertook the task of digitising the original book and editing, redesigning and laying out the revised edition, this publication would never have been.

2004

INTRODUCTION

The book *North Stradbroke Island* was first published in 1981 by Stradbroke Island Management Organisation (SIMO) – and became immediately known as 'the jellyfish book'. The organisation felt that there was a need to promote an awareness of Stradbroke Island's exceptional qualities. It was hoped that an understanding of the social and natural environment would be a guide to a planning process directed at retaining the island's unique character. The publication was achieved without outside financial help. SIMO members contributed to costs in proportion to their pocket. Within six months, books sales had cleared all costs and members were reimbursed.

A proposal to construct a bridge from the mainland to Stradbroke had met with much opposition. In 1978, island residents and regular visitors formed SIMO to fight the proposal. It was plain that a bridge would urbanise the island. However, after the Minister R J Hinze stated that 'such a proposal is not economically feasible under current circumstances',[49] the idea of a bridge was dropped in 1988. With the bridge abandoned, transport companies were prepared to invest in fast water taxi and vehicular ferry services from Cleveland to Dunwich. These services radically reduced Stradbroke's island isolation. With Stradbroke only 30 kilometres from densely populated south-east Queensland, the island provided a ready recreation resource that generated tourism.

Because major changes have drastically modified both the natural and social island environment, SIMO has revised and updated its 1981 publication. While sand-mining has provided island roads and jobs, boosting the island economy, and minerals for modern technology, the industry has disrupted the island's landscape and vegetation patterns. Unfortunately, beaches are no longer places for passive recreation, as 4 wheel drives treat the beaches as highways. Redland Shire Council has a permit to extract 53ML/day for island water to be pumped to the mainland; yet the impact of this rate of extraction on the island's wetlands is unknown. Native Title on Stradbroke is a pressing issue. As the State is the subdivider of island land, and because all Crown land is subject to Native Title, no further land can be opened up. Unrealistic land values have put real estate prices beyond the means of most. This is changing the island's social structure.

Despite major changes, Stradbroke is still a remarkable place and is well-known in scientific terms. The Royal Society of Queensland published in 1975 papers from a symposium held at Point Lookout and again in 1984 papers from a second symposium. Since then, new information published in scientific journals has been added to existing information on the island. It is one of the few places in Queensland that is well-documented and scientifically planned for. Unfortunately, the major plans that relate to Stradbroke's future are purely advisory; except in the case of the Point Lookout Development Control Plan, neither local nor State authorities are bound by them. Nearly 25 years after we wrote in the first edition of this book: 'The future of the island is by no means secure', Stradbroke is still not being well cared for. Redland Shire Council planning does not find favour with those who care about retaining the island's exceptional qualities. Over the 50-odd years that Stradbroke has been subjected to sandmining, the Department of Natural Resources and Mines has chosen to overlook breaches of mining conditions. We hold the view that not nearly enough land has been protected in the National Parks on the island and that Heritage-listed Point Lookout foreshore and Ramsar sites are far from properly managed. The international Ramsar agreement is of importance to the survival of migratory wading birds that each year fly from Alaska, China and Siberia to Moreton Bay.

Since 1978, SIMO has been an active body fighting to maintain the island's special qualities. Stradbroke remains an extensive wallum area that is accessible to Brisbane. Its flora and fauna are diverse and scientifically significant. It has unequalled educational potential. SIMO hopes that future planning will acknowledge and safeguard these qualities.

Those with conflicting interests about land use on the island will continue to put their arguments. Advocates for a bridge saw it as a means to promote the island for maximum benefit and the common good, but today – with rapid growth in southeast Queensland – there is greater awareness of the value of protecting Stradbroke as a rare 'natural' place close to Brisbane.

Despite the problems and conflicts over Stradbroke's future, the island remains a delightful, exciting and ever-changing place that rewards locals and visitors with its quiet pleasures and natural beauty.

The Natural Environment

Plate 1 Recent generations will go down in history as 'the greatest and most destructive vandals of all time'.[53] *As individuals, we can make a conscious decision to direct our activites to minimising environmental impact. (Authors)*

'Much of the surface of the earth supports plant and animal species. The distribution of these species is not at random, but is highly organised into complex patterns of intergrading variation'.[99]

Stradbroke has relatively uniform sandy soils low in nutrients, but its landscapes and plant and animal life are very diverse because of a high degree of micro-habitat diversity, abundant rain, and an equable climate. Sixteen different plant communities are recognised on the island. These are made up of nearly 500 species of plants, and are home to about 330 vertebrate species and many more, but poorly-known, invertebrate species. Island ecosystems are known to be very fragile because of their small size and isolation. Their flora and fauna are thus particularly vulnerable to extinction.

In the 180 or so years since European use of Stradbroke began, the island has been altered dramatically. Large tracts of wallum forest and swamp areas have been cleared for sandmining. A stand of the Swamp Orchid (*Phaius australis*) has been destroyed near Amity; patches of the fern (*Todea barbara*) which grows along swamp edges, and the Water-loving Daisy (*Olearia hygrophilia*), found only on Stradbroke, are under threat; and mangroves are constantly being destroyed or encroached upon. The rainforest at Myora has gradually been altered by human use, and introduced plants like Groundsel, Asparagus Fern (*Protasparagus densiflorus*), Peppermint Bush (*Schinus terebinthifolia*), Umbrella trees (*Brassaia aetinophyila*) and Lantana have spread island-wide. Additional pressures are inflicted on native animals by introduced cats, dogs, cane toads, rats and foxes.

Isolation of animal populations from their ancestral mainland stock on islands often leads to the evolution of new, peculiarly island species or forms. The dark brown, heavily-built Swamp Wallaby (*Wallabia bicolor*) is probably the most conspicuous native vertebrate on North Stradbroke Island. The species occurs widely on mainland eastern Australia from Victoria to northern Queensland. Populations of this species on the island today are certainly descendants of migrants from the present mainland area. Over a long period in an isolated environment free of many of the pressures on mainland populations, a golden-coloured variant of this wallaby has evolved and survived on North and South Stradbroke Islands. This rich, pure golden-coloured form has not been recorded anywhere else amongst Swamp Wallaby populations.

Plate 2 North Stradbroke Island is composed almost entirely of sand. Rocks are exposed at Dunwich, Point Lookout and the quarry at Main Beach. The island has been linked to the adjoining mainland in both recent and geological times. The plants and animals of Stradbroke reflect this close affinity. (Department of Mapping and Surveying)

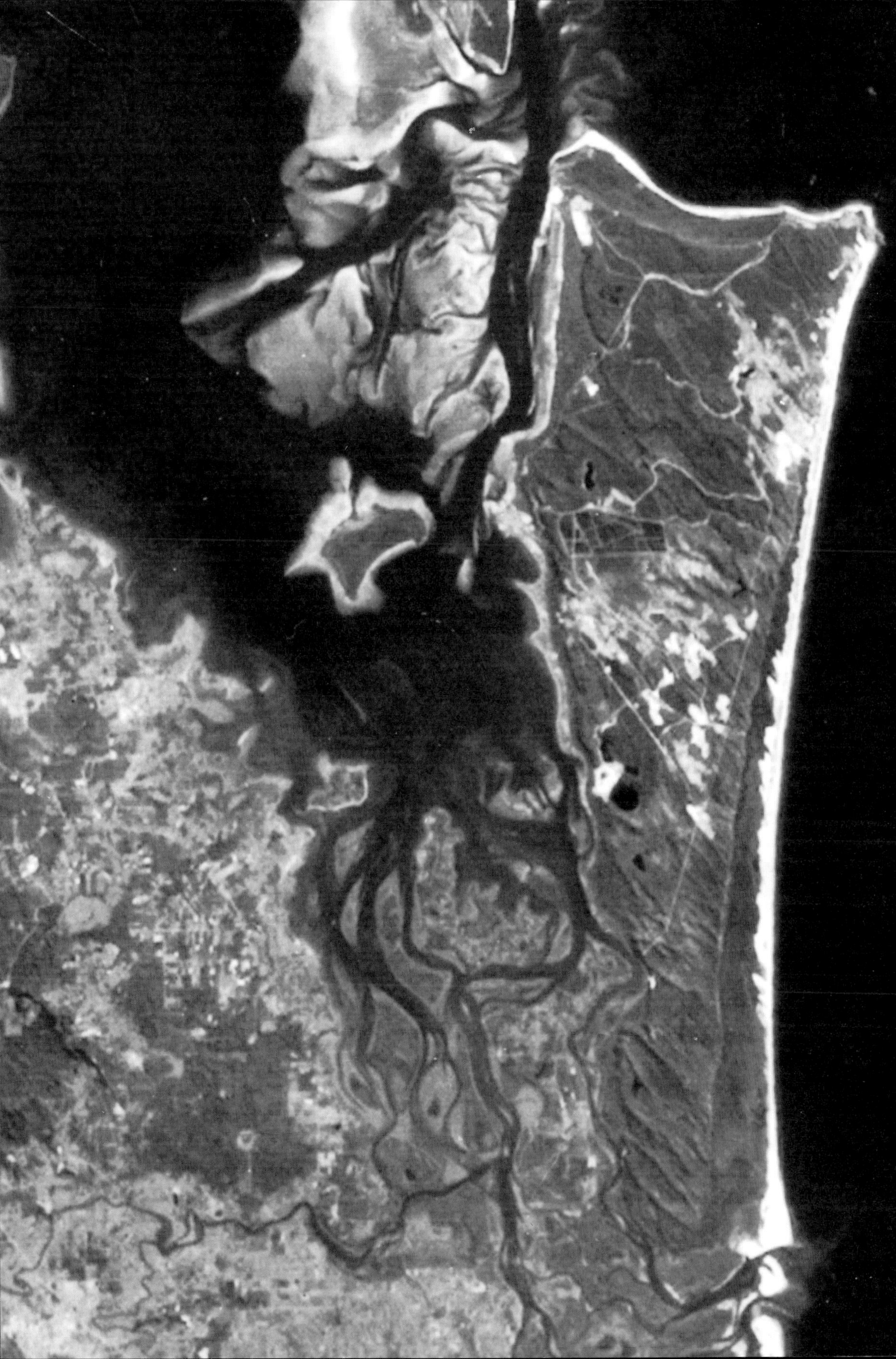

Islands are also known to act as refuges for species which become 'relicts' when their mainland source populations become extinct through pressures which do not affect the island populations. The Agile Wallaby (*Macropus agilis*) occurs on Peel and Stradbroke Islands. It presently occurs nowhere else in southern Queensland, being a coastal tropical species found north of Rockhampton in Queensland and in the Northern Territory. However, a colony of Agile Wallabies discovered in the Gold Coast hinterland is being assessed to establish if the wallabies are derived from mainland stock or from some unrelated captive species. *Macropus agilis* are very common elements of recent (two million years old) fossil remains from southeastern Queensland sites. What is obvious is that *Macropus agilis* was once common on the mainland in southeastern Queensland; that it colonised what are now North Stradbroke and Peel Islands during one of the low sea level periods when they were connected; and that the source populations became extinct on the mainland through some pressure that did not occur on Peel and Stradbroke Islands. *Macropus agilis* continues to survive on Peel Island, but traffic and domestic dogs appear to have extinguished the Stradbroke colony.

North Stradbroke is a land bridge island. It was separated from what is presently South Stradbroke Island in 1896, when a violent storm caused the break now known as Jumpinpin.[105] The old 'Stradbroke Island' (= the present North and South Stradbroke Islands) may also have been joined to The Spit at Southport until recently. This break was first reported in 1823 by John Oxley. Neither of the two great navigator-explorers, Cook (1770) or Flinders (1799), observed it. Considering the minute detail in which other parts of the coast were described by Cook and Flinders, it is reasonable to assume that they would have reported the break if it existed. It seems likely, therefore, that the present Southport Spit, South Stradbroke Island, and North Stradbroke Island were all connected until fairly recently, forming a long, narrow sand spit. (See Appendix VI.)

In recent geological time (the last two million years), the connections between what is presently North Stradbroke and the adjoining mainland have been even greater than the narrow spit connection during the last 200 years. A drop of only 28 metres in present sea level would link wallum areas on the mainland and the islands of Moreton Bay. Sea levels have fluctuated dramatically even in the last two million years and there

Plate 3 Cabbage Tree Palms (Livistona australis) form dense stands at the southern end of the Eighteen Mile Swamp. (Authors)

Plate 4 Blackbutt (Eucalyptus pilularis) and Lilly-pilly (Syzygium spp) in a secluded gully on the old road to Blue Lake. Such protected pockets are moister than surrounding areas and can thus support denser vegetation, some of which suggests the early stages of rainforest development. (Authors)

Fig 1 Fruit of one of the few rainforest plants on the island, Lilly-pilly (Syzygium sp)

have been many opportunities for plants and animals to colonise the area now called North Stradbroke Island. The many mainland-North Stradbroke links which have occurred account for the similarities between the present fauna and flora of each area. Alternating separations by sea, with consequent isolation of the island flora and fauna for long periods, account for the differences between them. Because North Stradbroke Island has been a small area isolated by sea from the adjoining mainland many times during its geological history, it has fewer habitats than the mainland and, as expected, reduced faunal diversity. Many species present on the adjoining mainland are not present on Stradbroke. Rainforest species are scarce, for example, because rainforest is not extensive on the island. The few rainforest animals on North Stradbroke Island occur in the dense moist forest at Myora Springs and in the closed palm forest at the southern end of the Eighteen Mile Swamp.

Most, if not all, plant species that occur on the island also occur in wallum habitats on the adjoining mainland. Many are uniquely adapted to life under special conditions – acidic swamps, exposed and wind-blown slopes, salt-laden winds, and sands low in nutrients. Trees and shrubs on high, dry, exposed dunes are sheared off and stunted by the stresses of water shortages and strong southeasterly winds. The same species grow taller and straighter in moister, more sheltered positions in gullies. In a few very secluded and moist depressions, these taller forests suggest the beginnings of rainforests with Blackbutt (*Eucalyptus pilularis*) and Lilly-pilly trees (*Syzygium* spp). Some rare species survive on Stradbroke. The large Swamp Orchid (*Phaius australis*) occurs in isolated pockets and once was common on the mainland, before development and landfill pressures. While Christmas Bells (*Blandifordia grandiflora*) are still found in certain mainland areas, the small patch which grew at Yerrol Creek on Stradbroke no longer exists. This is likely due to excessive picking of blooms and too frequent bush fires.

The Daisy, *Olearia hygrophila*, is particularly interesting. It is an extremely rare plant. While collecting specimens in the Moreton Bay District for the Royal Botanical Gardens, Kew, Allan Cunningham recorded the species in a swamp on Stradbroke in 1828. It was recorded once more, in 1888, but then 'lost' for close to 100 years, until found again on the island in 1974 by Mr Selwyn Everist, the Queensland Government Botanist. It is not known from any other localities.

Plate 5 Blue Lake is fed from the island's underground water. On sunny days the lake's water reflects the deep blue of the sky. It has a maximum depth of 9.4 metres. (Bruce Cowell)

Flora

Wetlands and lakes: Swamps fringe the island's northern, eastern, and southern shores, and many lakes and swamps have formed in inland depressions. Some of these, like Blue Lake, are water table lakes fed from the vast underground water supply. Several (for example Brown Lake) are 'perched' above the water table in depressions lined with accumulated vegetation which forms an impervious layer to hold water.

'The Eighteen Mile Swamp is particularly remarkable. This great linear trough in the sand, lying seaward of the truncated giant dunes of the main sand mass, and kept full of water by seepage from the sand mass rather than by surface run-off, is unique. The only feature resembling it in the great sand masses further north are the small swamps at Peregian and these are only midgets compared to the Eighteen Mile Swamp'.[38]

Sword Grass (*Gahnia sieberana*) fringes and ranges across most swamps. Paperbarks (*Melaleuca quinquenervia*) and Swamp Mahogany (*Eucalyptus robusta*) are the dominant trees in the swamps.

Plate 6 Tea-Tree Lagoon, one of the island's many less accessible lakes (Authors)

Unfortunately the Eighteen Mile Swamp is threatened by the mining of the adjacent high dunes which form Consolidated Rutile's Enterprise mining lease. Redland Shire Council's pumping water from the swamp for the mainland, and Consolidated Rutile's extracting water for its mining operations, are additional threats.

Frontal dune communities: Vegetation traps sand and stabilises the foredunes which form the buffer zone between the beach and the land. Plants which can grow on exposed seaward slopes have to be salt-tolerant and ground-hugging. Some are succulents. Beach Spinifex (*Spinifex hirsutus*) is the dominant plant. The creepers, Goat's Foot (*Ipomoea pes-caprae*), with its mauve flowers, and the succulent triangular-leafed Pigface (*Carpobrotus glaucescens*) which has deep purple and lemon-white flowers, add a splash of colour to the white dunes. The Snake Vine or Twining Guinea Flower (*Hibbertia scandens*) produces a fetid-smelling bright yellow flower.

Foredune stability in the face of almost continuous strong wind is dependent on retention of covering vegetation. If the vegetation is disturbed or cleared, dunes quickly become mobile, creeping over adjoining areas. Vegetated dunes act as sand reservoirs. Once destroyed, they can no longer replenish sand to scoured beaches.

Fig 2 Guinea Flower (Hibbertia scandens)

Plate 7 Blue Lake National Park (The late Bob Pattison)

The Screw Palm (*Pandanus pedunculatus*), the fruit and young leaves of which are edible, and the Coast Sheoak (*Casuarina equisetifolia*) grow on exposed dunes and headlands. Coast Banksia (*Banksia integrifolia*) is often stunted and grotesquely-shaped from exposure to constant strong winds. The large-leafed, yellow-flowering Cotton Tree (*Hibiscus tiliaceus*) is less tolerant. This sprawling tree often covers a large area, but grows only in dune depressions and on less exposed beaches.

Fig 3 Goat's Foot (Ipomoea pescaprae)

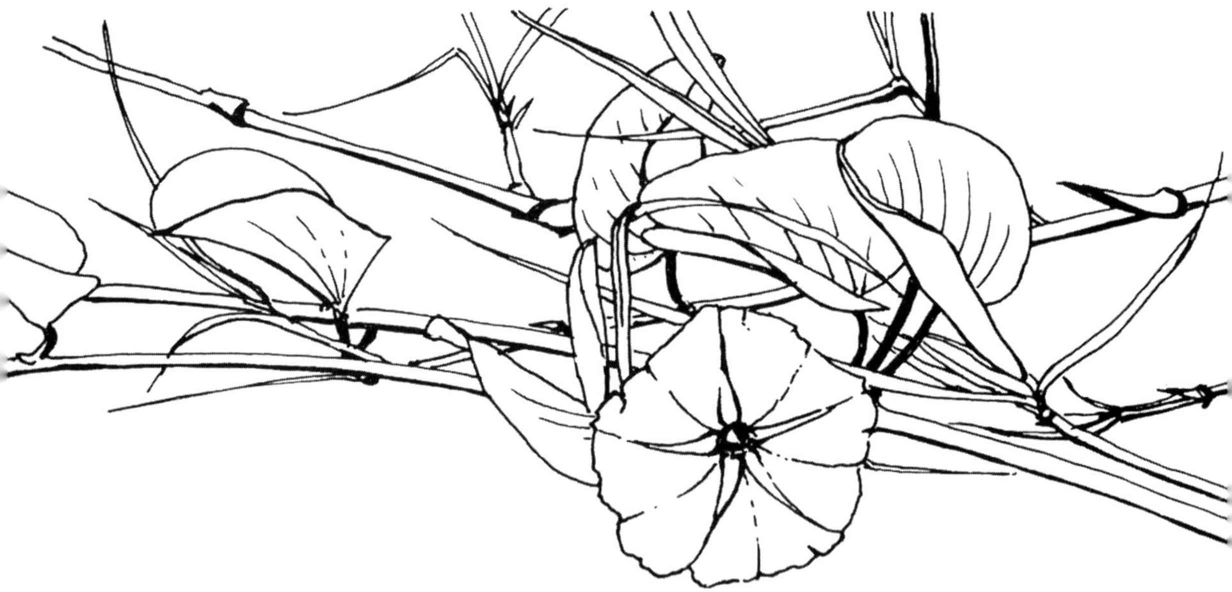

Plate 8 Beach Spinifex (Spinifex hirsutus) in seed. This species helps trap and stabilise sand on the island's foredunes. (Authors)

Plate 9 The hardy Screw Palm (Pandanus pedunculatus) and Sheoaks (Casuarina equisetifolia) thrive on the exposed headlands at Point Lookout. (Authors)

Plate 10 Crinum Lily (Crinum pendunculatum) on the headlands at Point Lookout (Authors)

Plate 11 Strong winds make life difficult for the Coast Banksia (Banksia integrifolia). Where this species occurs in more protected areas, it grows tall. (The late Bob Pattison)

Plate 12 Bull Banksia (Banksia serrata). On the exposed seaward faces of the high dune systems, all the vegetation is stunted. (Authors)

High dune systems: *Eucalyptus* species – Scribbly Gum (*E signata*), Blue Gum (*E tereticornis*) which is favoured by the island's koalas, Planchon's Stringybark (*E planchoniana*), Blackbutt (*E pilularis*) and Bloodwoods (*E intermedia* and *E gummifera*) – dominate the high dunes which are composed of infertile sands.

Sheoaks (*Casuarina torulosa* and *Casuarina littoralis*) tend to grow in single-species colonies. The name 'Sheoak' derives from the sighing sound made by the wind passing through the trees' needle-like branchlets, which are modified leaves. The leaf 'needles' form a thick springy mat on the forest floor. During winter, when the male *C torulosa* trees are covered with pollen, they become red-brown. The Cypress, like the Sheoak, is wind-pollinated. When the dark green foliage of the Cypress was dusted with rust pollen, the Aborigines referred to the event as 'The Smokes of the Cypress'. Away from the strong salt-laden winds, Bribie Cypresses (*Callitris columellaris* and *C rhomboidea*) grow tall and straight.

A variety of ground vegetation grows under the forest trees. In spring, the Bridal or Wedding Bush (*Ricinocarpos pinifolius*), with its masses of snow-white blooms, is spectacular. Blueberry Ash (*Elaeocarpus obovatus* and *E reticulatus*) are tall shrubs which bear cream and occasionally delicate pink pendulous flowers, followed by attractive blue berries.

Fig 4 Crinkle Bush (Lomatia silaifolia) 'responds to fire with an immediate outburst of growth followed by an unusually heavy flowering'.[89]

Plate 13 Blueberry Ash (Elaeocarpus reticulatus) of the open forest in flower in November (Authors)

Plate 14 Flowers of the Bull Banksia (Banksia serrata) (Authors)

Fig 5 One of the island's many May bushes (Leptospermum sp)

Fig 6 Wedding Bush (Ricinocarpos pinifolius)

The Crinkle Bush (*Lomatia silaifolia*) and a number of wild May species (*Leptospermum* spp) have conspicuous white flowers. Possum Boronia (*Boronia rosmarinifolia*) is a tufted pink-flowering low shrub; when crushed, its aromatic leaves give off an odour similar to a possum. The hardy Bracken Fern (*Ptendium esculentum*) is common in most understorey situations. Grass Trees (*Xanthorrhoea* spp) are specially adapted to thrive despite fires, which increase the number of spear-like inflorescences. The Aborigines believed that if the Grass Tree flower spears were cut, rain would follow. The base of the plant's inner leaves is edible and the trunk produces a resin.

Plate 15 Grass Trees (Xanthorrhoea sp) near the old Blue Lake road (Authors)

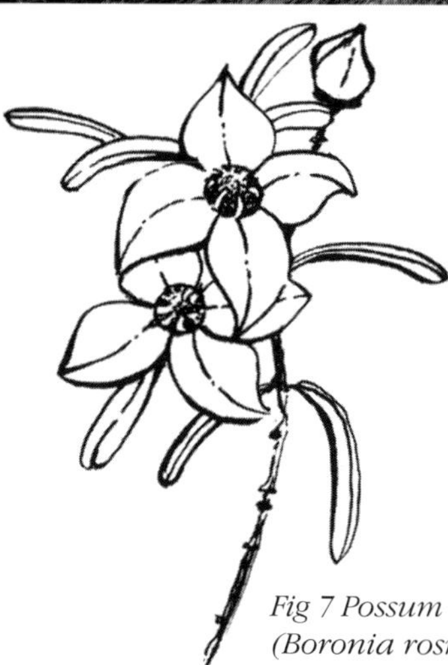

Fig 7 Possum Boronia (Boronia rosmarinifolia), a pink-flowered species from dry ridges

Fig 8 (below) Golden Pea Flower (Gompholobium sp), one of the common plants along roadsides

Fig 9 Bat's Wing Fern
(Histiopteris incisa)

Plate 16 Tangled masses of aerial roots anchor mangroves in soft mud. (Authors)

Plate 17 The salt-tolerant Golden Mangrove Fern (Acroistichum speciosum) is common on the fringes of mangrove forests and saltmarshes. (Authors)

Mangroves: The intertidal zones are colonised by mangroves. These are highly specialised plants capable of adapting to varying degrees of salinity. At high tide, the mangrove roots are completely inundated by salt-water. The familiar mangrove 'pegs' (pneumatophores) provide aeration for underground roots in poorly-drained silt mud. The Grey (or White) Mangrove (*Avicennia marina*) and the Small-stilted Mangrove (*Rhizophora stylosa*) are the most common species occurring on Stradbroke. They fringe the island's western shores and Swan Bay in the south. Large stands are accessible at Amity and Myora Springs.

Mangroves have built-in mechanisms to reduce internal salt accumulation. Some exude salt through their root systems; others secrete or extract salt through their leaves. Many have 'grotesque' aerial root systems which anchor the tree in soft mud, and also protect foreshores in storm surges.

'Overseas studies have shown that in some coastal waters, 80% of fish caught commercially were linked to food chains dependent ultimately on mangroves'.[61]

Nutrients derived from the decay of mangrove and associated litter provide food on which micro-organisms at the base of the food chain depend. As well as supporting a diverse marine fauna, mangroves are the home of some unusual terrestrial fauna. On Stradbroke, several birds (eg, the Mangrove Honeyeater, the Mangrove Kingfisher, the Mangrove Warbler), and the rare Water Mouse, previously referred to as the False Water Rat, are confined to mangrove habitats.

'There exists on all shores of mud or mixed mud and sand, a fauna which is not found elsewhere and on which certain elements demand the shelter of vegetation. In the tropics and sub-tropics these elements will always be associated with mangroves.'[64]

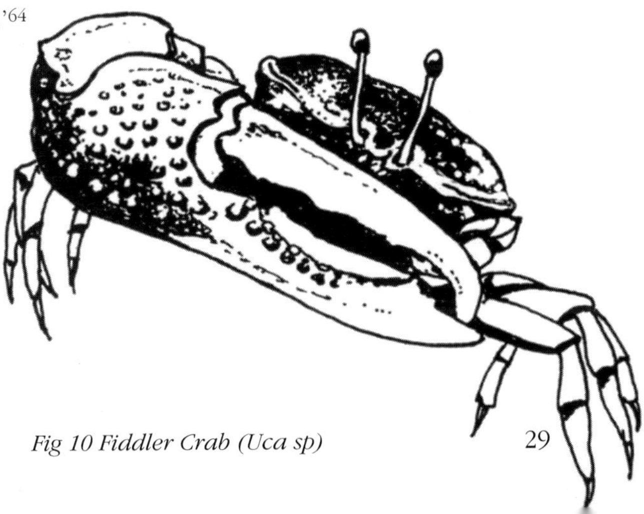

Fig 10 Fiddler Crab (Uca sp)

Plants of special interest: The carnivorous plants, ferns, orchids, and wattles which grow on Stradbroke are of special interest.

Three species of carnivorous plants (*Drosera* spp) can be found in moist, sheltered but sunny areas. They all catch tiny insects in sticky tentacles, an adaptation which enables them to supplement their nutrient supply. Even more astounding is the adaptation of a bug (Family Miridae) to live on these plants and 'steal' their food from the plants – its insect prey trapped in the sticky tentacles. These bugs have long legs which enable them to move about between the 'goo' of the tentacles to suck juices from the prey of the plants without being stuck and digested along with the rest.

With the exception of the Bracken Fern (*Pteridium esculentum*), which grows almost everywhere, and the wiry Coral Fern (*Lycopodium cernuum*) of sunny, moist areas, the ferns which grow on Stradbroke favour the damp, semi-shaded swamps and moist valleys or embankments. Soft-fronded Mountain Bracken (*Culcita dubia*) covers whole hillsides in some shaded pockets. The swamps support stands of King Fern (*Todea barbara*), Bat's Wing Fern (*Histiopteris incisa*) and the twining climber (*Lygodium microphyllum*); but the dominant species in these areas is the Bungwall Fern (*Blechnum indicum*), which has soft, reddish new fronds. These turn to dark green and become leathery with age.

Some of the orchids which grow on the island are spectacular. The Swamp Orchid (*Phaius australis*) is the largest ground orchid in Australia. Its flower spikes sometimes grow to almost 2 metres. In August each year these orchids flower along swamp edges. The pink-brown and yellow flowers were once a common sight, but clusters of these plants are now quite rare owing to destruction of their habitat by road construction and sandmining, and also through collection by misguided orchid fanciers. Small (about 20 cm) pink and white *Caladenia* grow in forest leaf litter in shady areas. Another small species, the Flying Duck Orchid (*Caleana major*) prefers more sun. Most plants have leaves through which they utilise the sun's energy to sustain life. Some orchids lack leaves so have another means of tapping the sun's energy. Two of these, the Climbing Orchid (*Galeola cassythoides*) and the Hyacinth Orchid (*Dipodium* sp), occur on the island. Both depend for their survival on a root fungus. Both flower in spring. *Galeola cassythoides* produces clumps of delicately perfumed brown-yellow blossoms from a vine that looks lifeless; *Dipodium* has spotted mauve or yellow flowers on what looks like a plain grass stalk.

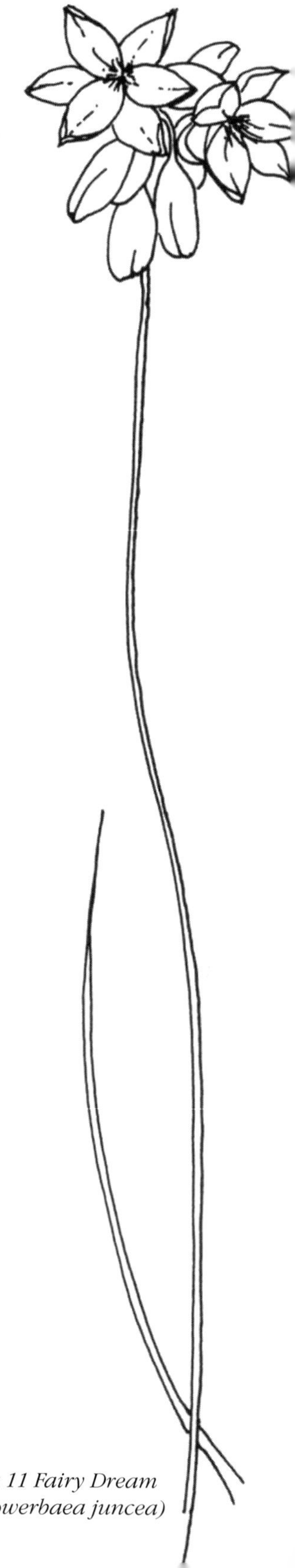

Fig 11 Fairy Dream (Sowerbaea juncea)

Plate 18 '...each tentacle secretes in the gland at its tip a digestive juice. Over a period of days the soluble matter of its prey is reduced to a fluid that is absorbed by the plant. The tentacles then assume their first position, fanned out and gluey tipped, ready to trap the unwary. The dry husks of the earlier victims drop lightly from the leaf and the table is set for another meal.'[36] Drosera peltata is one of the carnivorous plants on the island. (Bruce Cowell)

Wattles (*Acacia* spp) are important colonisers of high dunes after they have been disturbed either naturally (eg, by erosion) or by humans (following mining or road building). The Brisbane Black Wattle (*Acacia concurrens*), which has lamb's tail-like bright yellow flower tassels, is the dominant species. It forms dense thickets in some areas, following clearing. Sweet Wattle (*Acacia suaveolens*) and Prickly Moses (*Acacia ulicifolia*) are also common trees.

Fig 12 Boronia falcifolia, a common plant at Brown Lake, has purple flowers in spring.

Plate 19 The Swamp Orchid (Phaius australis) was once common in swamps on the island. Now it occurs only in isolated patches. Road construction, sandmining, and misguided orchid fanciers have all taken their toll. All native plants are now protected under State Nature Conservation legislation. The Phaius is totally protected. (Authors)

Fig 13 Boronia safrolifera is widespread on the island. Its spring flowers are pink.

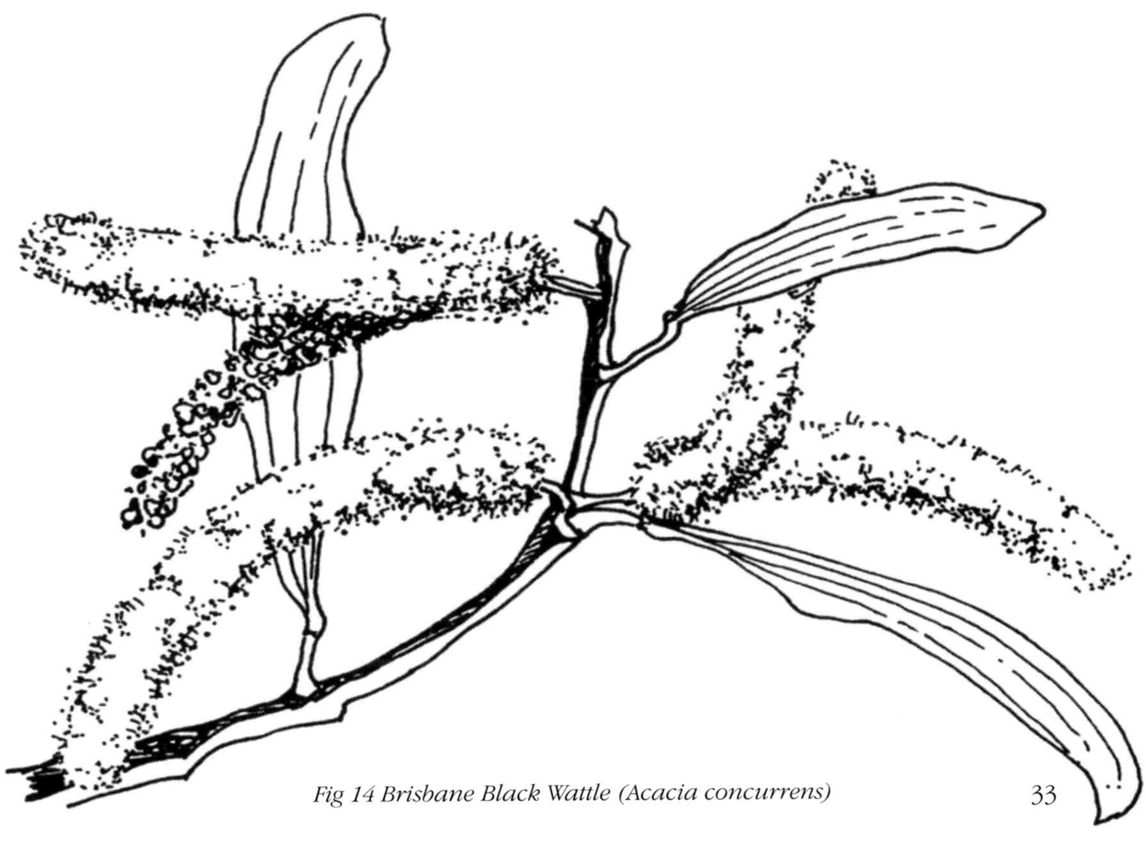

Fig 14 Brisbane Black Wattle (Acacia concurrens)

Plate 20 Ferns thrive in secluded, moist pockets. (Bruce Cowell)

Plate 21 Bungwall (Blechnum indicum) is probably the most common fern growing in swamps on North Stradbroke Island. (Authors)

Fig 15 Climbing Orchid (Galeola cassythoides)

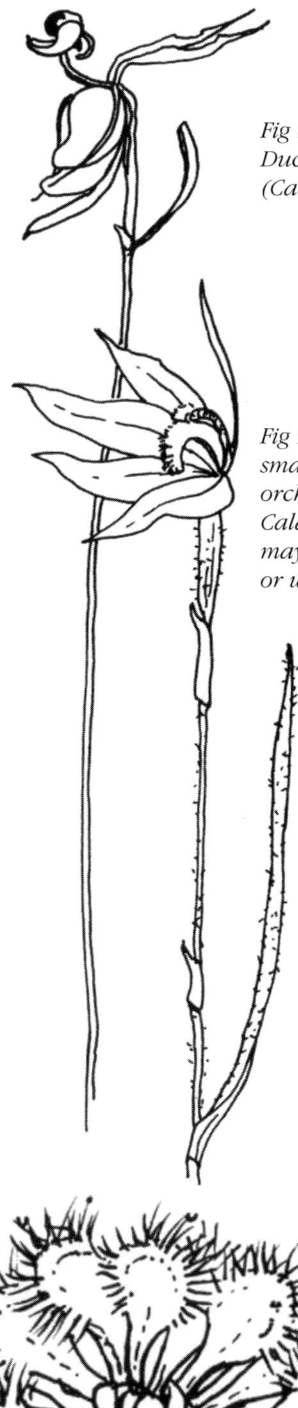

Fig 17 Flying Duck Orchid (Caleana major)

Fig 18 Many small species of orchids occur. Caladenia sp may have pink or white flowers.

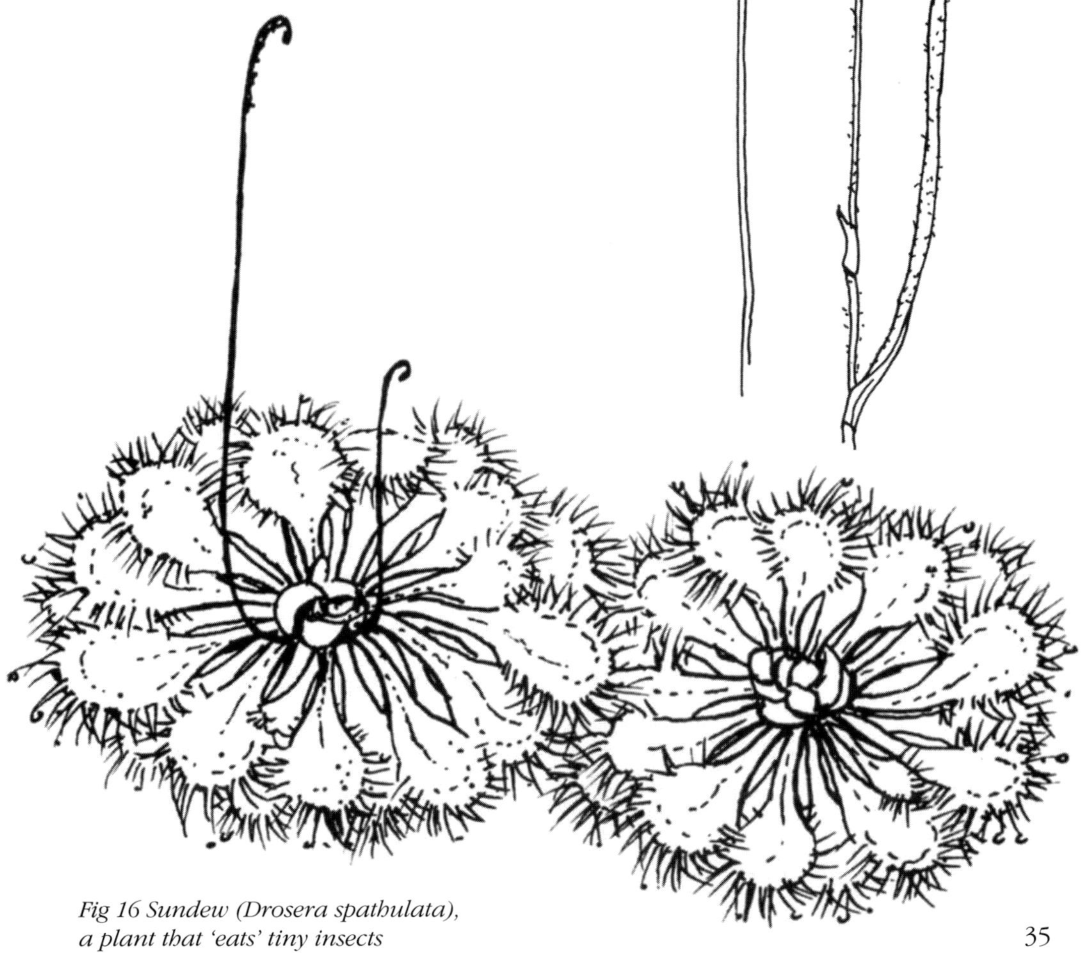

Fig 16 Sundew (Drosera spathulata), a plant that 'eats' tiny insects

Plate 22 (opposite)
Swamp Harriers (Circus approximans) are uncommon on Stradbroke. They hunt their prey, small birds like waders and young ducks, by flushing them from cover and then swooping down on them. (Queensland Museum)

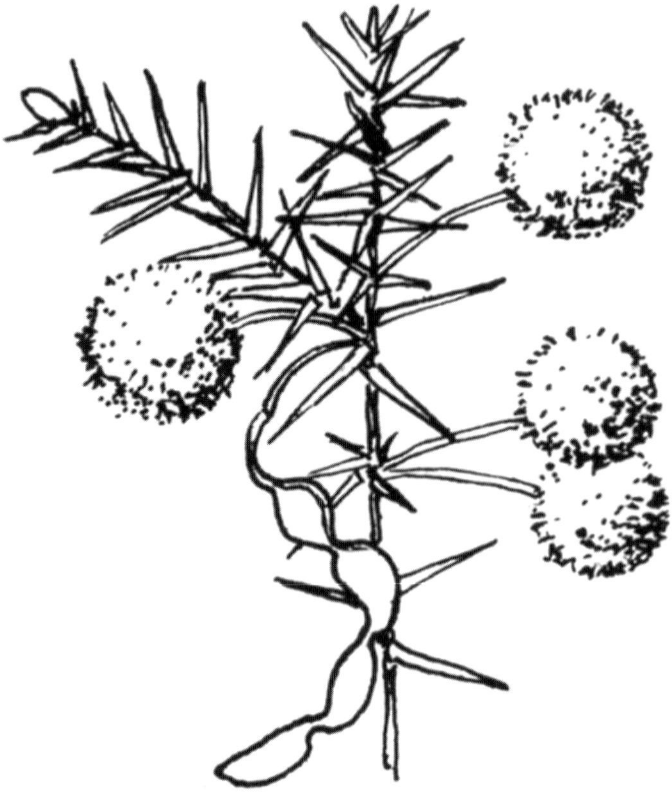

Fig 19 Prickly Moses (Acacia ulcifolia)

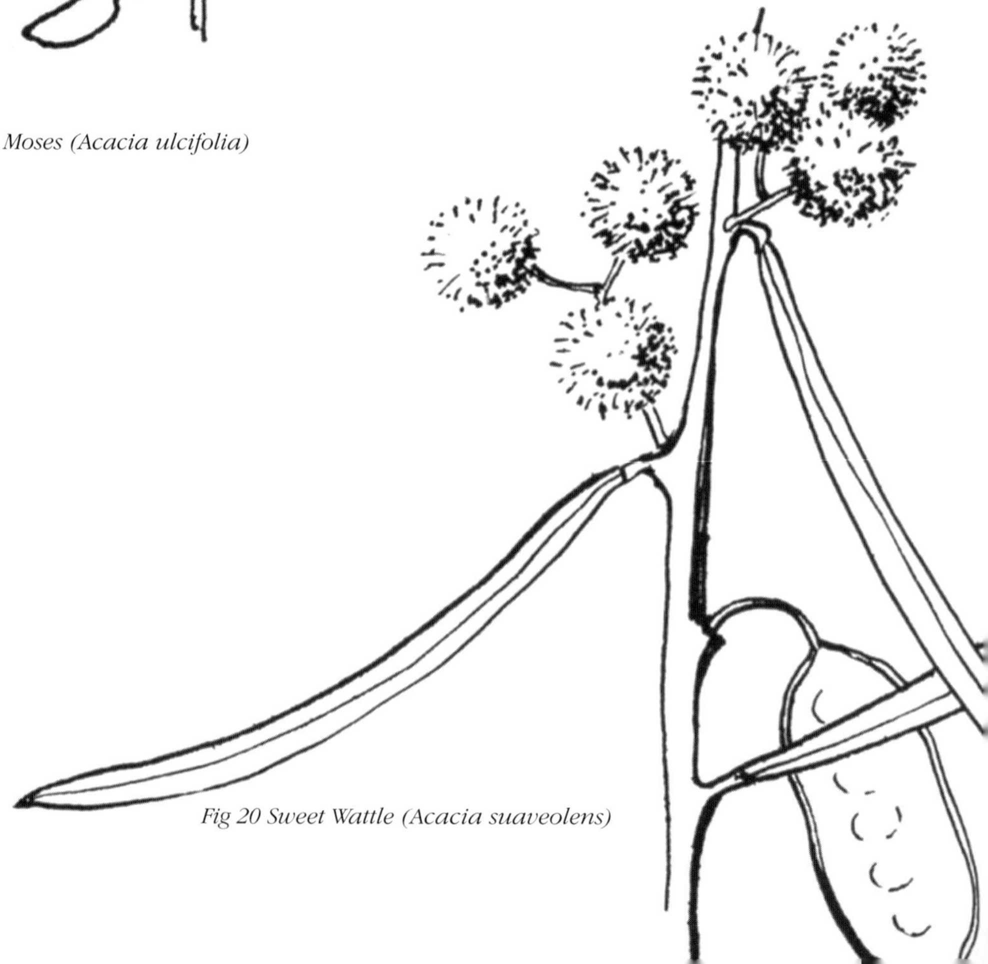

Fig 20 Sweet Wattle (Acacia suaveolens)

Vertebrate Fauna

Plate 23 Black Duck (Anas superciliosa) are a common sight on Stradbroke's many lagoons and lakes. Their young are not easy to see, however. Black Ducks feed on aquatic vegetation. (Queensland Museum)

Land animals: Nineteen species of frogs (including the introduced Cane Toad), 51 species of reptiles, more than 250 species of birds, and 45 species of mammals have been recorded from the island. This is a very diverse fauna. (Species are listed in Appendix I.) Since the last lists were compiled in 1975, many new records have been added by the Queensland Museum and the Environmental Protection Agency (National Parks and Wildlife). These additions are included here.

Many of Stradbroke's vertebrate animals are rare or uncommon or occur in a limited south-east Queensland range that includes Stradbroke Island. Populations of most species are small. This, combined with their shy, secretive natures, means that only a handful of species are encountered. Many of the larger species are nocturnal and these are especially vulnerable to road traffic.

The following system of ranking species according to their biological significance and conservation status has been devised to facilitate comparisons of the fauna from different areas being subjected to change and in which, therefore, habitat and faunal destruction may take place.

Rare species: rare species of varying outstanding biological or phylogenetic significance, seldom recorded and poorly represented in reference collections;

Species of limited Australian range: species of limited range and of biological and/or biogeographic significance, the Moreton Region representing a significant part of their range;

Uncommon species: species whose occurrence in the Moreton Region represents the southern, northern or eastern extent of their range.[67]

On this basis, 17 vertebrate species of North Stradbroke Island rate as being important biologically. Of particular interest is the occurrence on the island of the Agile Wallaby (*Macropus agilis*), False Water Rat,

Plate 24 Swamp Wallaby or 'Stinker' (Wallabia bicolor) (Queensland Museum)

now referred to as Water Mouse (*Xeromys myoides*), the skink (*Anomalopus truncatus*), and the 'acid' frogs (*Litoria cooloolensis*, *L olongburensis*, *L freycineti*, *Ranidella tinnula*); and the Golden Stradbroke Wallaby (a colour phase of the Swamp Wallaby) is unique to North and South Stradbroke Islands.

The Agile Wallaby is a relict species on Stradbroke and Peel Islands. It is now essentially a northern species, occurring across much of coastal Australia north of Rockhampton. In recent geological times, however (the last two million years), it was a common species in mainland southeastern Queensland; here, owing to pressures which apparently did not prevail on Stradbroke and Peel, the Agile Wallabies became extinct. But now the Agile Wallaby is apparently extinct also on North Stradbroke Island, presumably as a result of human pressures.

The False Water Rat, now known as the Water Mouse, is one of the rarest mammals in Australia. It was found almost by accident by American students working in the mangrove forests at Myora in 1979. Subsequent Queensland Museum research on the North Stradbroke population revealed the species' dependence on the preservation of mangroves and the abutting super-littoral zone for foraging and nesting. Its continued abundance on the island's western coast is dependent on the preservation of the integrity of the island's richly diverse mangrove community. Elsewhere, the species is known from a handful of localities in southeastern and mid-eastern Queensland, coastal Northern Territory and southern Papua New Guinea.

Plate 25 Swamp Wallaby or 'Stinker', a victim of roadkill (Authors)

The Legless Skink (*Ophioscincus truncatus*) has a very restricted distribution. It is not found outside southeastern Queensland where it is known from only a few rainforest and wallum areas, including Stradbroke.

The 'acid' frogs are of special interest to scientists and students. These frogs can breed in waters of low pH, an adaptation which has resulted in their recognition as a discrete group ecologically. North Stradbroke Island is one of the best areas to observe these species because of its accessibility, and because its swamps, the frogs' habitats, are also accessible.

The Golden Stradbroke Wallaby: In 1922 Heber Longman, who was Director of the Queensland Museum, described as new a golden-coloured wallaby from North Stradbroke. He named it *Wallabia welsbyi* after Tom Welsby who had a long association with both Moreton Bay and the island. For many years biologists were curious about the status of this animal which had been seen by only a few people on both North and South Stradbroke Islands. Was it a 'good' species distinct from its mainland relatives or just a colour form of the common 'Stinker' or Swamp Wallaby (*W bicolor*)? Research by Jim McEvoy[70] of the National Parks and Wildlife Service has shown conclusively that *W welsbyi* is an unusual colour variety of *W bicolor* restricted to North and South Stradbroke Islands.

The Golden Wallaby is a spectacular animal. Seeing one is a never-to-be-forgotten experience. The few specimens that have been seen have been observed along road sides, usually late in the afternoon or at night when wallabies come out to feed.

Right: Plate 27 Bush Stone Curlews (Burhinus magnirostris) are masters of disguise. They favour open-forested areas on Stradbroke. Nesting pairs are not uncommon in the towns of Dunwich, Amity and Point Lookout. (Queensland Museum)

Lower right: Plate 28 Frogmouths (Podargus strioides) are very common on the island but they are not seen often because they are so well camouflaged and they hunt at night. They eat insects and any mammals they can catch. (Queensland Museum)

Below: Plate 26 Purple Crowned Pigeons (Ptilinopus superbus) are rare on Stradbroke. They have been seen in the Adder Rock camping ground and in the rainforest at Myora. They feed on fruit. (Queensland Museum)

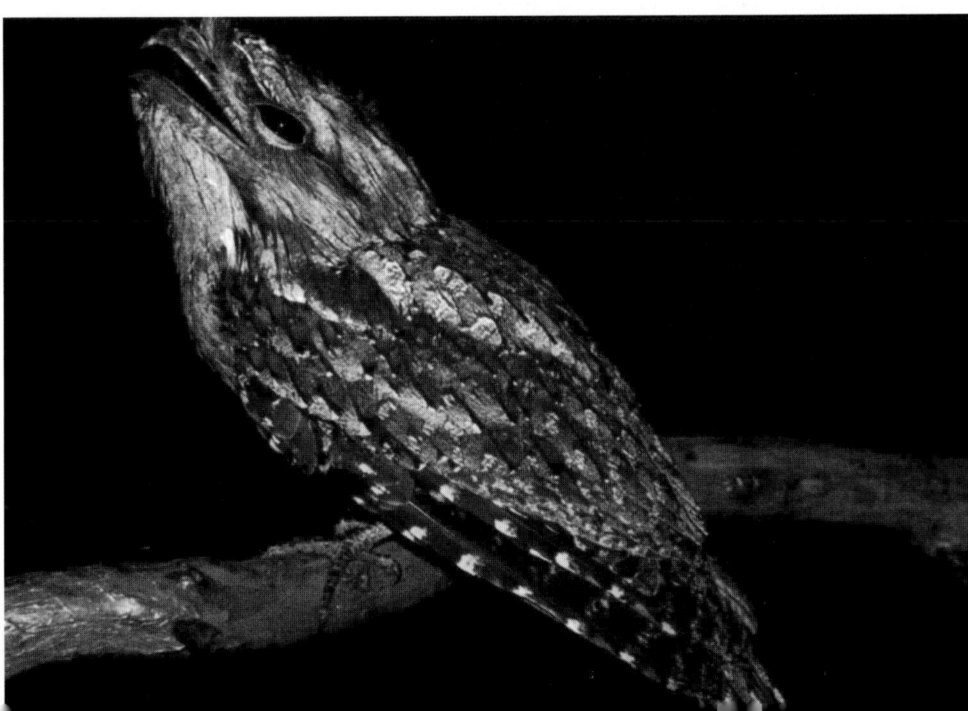

Plate 29 Striped Skink (Ctenotus taeniolatus) is a common, fast-moving lizard on the island. (Queensland Museum)

Marine reptiles and mammals: Green and Loggerhead Turtles (*Chelonia mydas*, *Caretta caretta*) are a common sight, and Leatherbacks (*Dermochelys coriacea*) are seen infrequently in the waters round Stradbroke. Five individual Loggerheads make between ten and 20 nests each year on the ocean beaches, and Green Turtles nest on the island irregularly. Sea snakes are frequently found washed up on Main Beach.

Humpback Whales (*Megaptera novaeangliae*) are seen from Point Lookout as they migrate north from Antarctic waters to breed in warmer reef waters. Most north-bound sightings occur between May and July. After calving, the whales and their young return south to their Antarctic feeding grounds. Herds moving south are seen from Point Lookout between August and October. Humpbacks are filter feeders. They lack teeth. Instead, they have baleen-fringed plates hanging from the upper jawline with which they sieve small invertebrates, mainly krill, which form massive concentrations in Antarctic waters. In 1952 a whaling station was established at Tangalooma, Moreton Island, to process the Humpbacks hunted on their annual migrations. By the early 1960s, Humpback populations had declined so dramatically that quotas could be filled only with difficulty. The Tangalooma station closed in 1962.[12, 79] Since then, populations have rapidly increased at an annual rate of around 12%.

Small whales or dolphins are common in Stradbroke's bay and ocean waters. The Bottlenose Dolphin (*Tursiops truncatus*) is the most common species.

Dugongs (*Dugong dugon*) are once again secure in protected bay waters near Stradbroke. Moreton Bay is one of the most important Dugong areas in the world. At least 700 Dugong are known from the area, which is home to the southernmost population in Australia. The herds are totally dependent on the seagrasses growing between Tangalooma Point, Moreton Island and Peel Island. The Amity Banks (just to the northwest of Amity Point) and those just southwest of Reeders Point (Moreton Island) are the richest feeding grounds.

Dangerous snakes: Sixteen species of snakes have been recorded on the island. Three species – the Brown Snake (*Pseudonaja textilis*), the Death Adder (*Acanthophis antarcticus*), the Tiger Snake (*Notechis scutatus*) – are always capable of inflicting a fatal bite. The Red-bellied Black Snake (*Pseudechis porphyriacus*), the Rough-scaled Snake (*Tropidechis carinatus*), and the Small-eyed Snake (*Cryptophis nigrescens*) are capable of inflicting a fatal bite in special circumstances (eg, if a large snake bites a small child). Bites from specimens of Stephen's Banded Snake (*Hoplocephalus stephensii*) and the Yellow-faced Whip Snake (*Demansia psammophis*) should also be treated with caution because they sometimes result in severe local symptoms. To find half of the species on the island capable of inflicting bites varying from fairly to very dangerous comes as a rather nasty shock. It should always be remembered, however, that island populations of these snakes are low and that snakes are encountered only infrequently anyway, because of their shyness. Large dangerous species are rare on the island. The Tiger Snake, for example, is known from only one museum record, and the Brown Snake, a common species on the adjoining mainland, from only four museum records for the island.

Plate 30 The Death Adder (Acanthophis antarcticus), which is usually nocturnal, is one of Australia's most dangerous snakes. Specimens have been found in many areas and it seems that they occur all over the island. Known as Moo-loo-kul to the Stradbroke Aborigines, the Death Adder is now uncommon on the island, probably because of the spread of Cane Toads. Adders see all frogs as food (see page 132). (Queensland Museum)

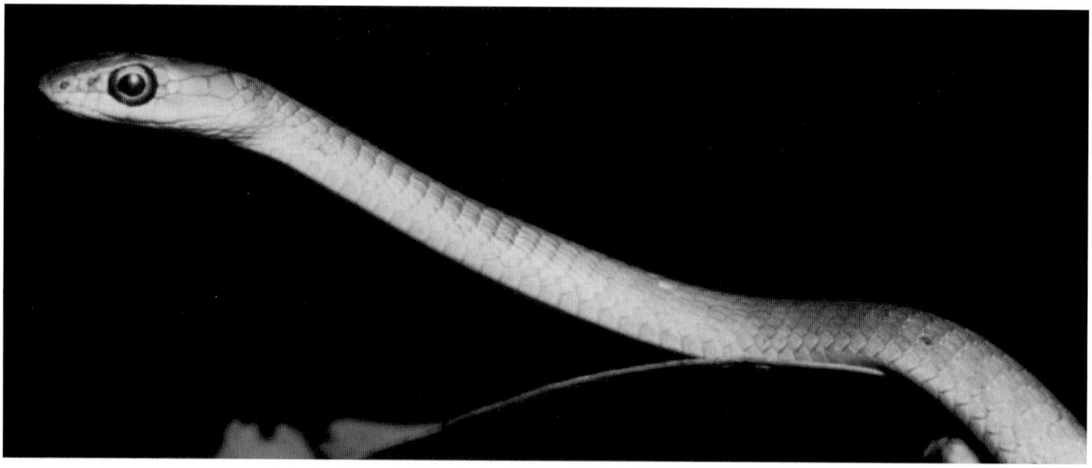

Plate 31 The Green Tree Snake (Dendrelaphis punctulatus) occurs widely on the island and is probably the most common snake near settlements. It is a slender, elegant snake, often with a canary-yellow throat. Green Tree Snakes have no venom so they are completely harmless. They feed mainly on frogs. If they ingest Cane Toads, they die. (Queensland Museum)

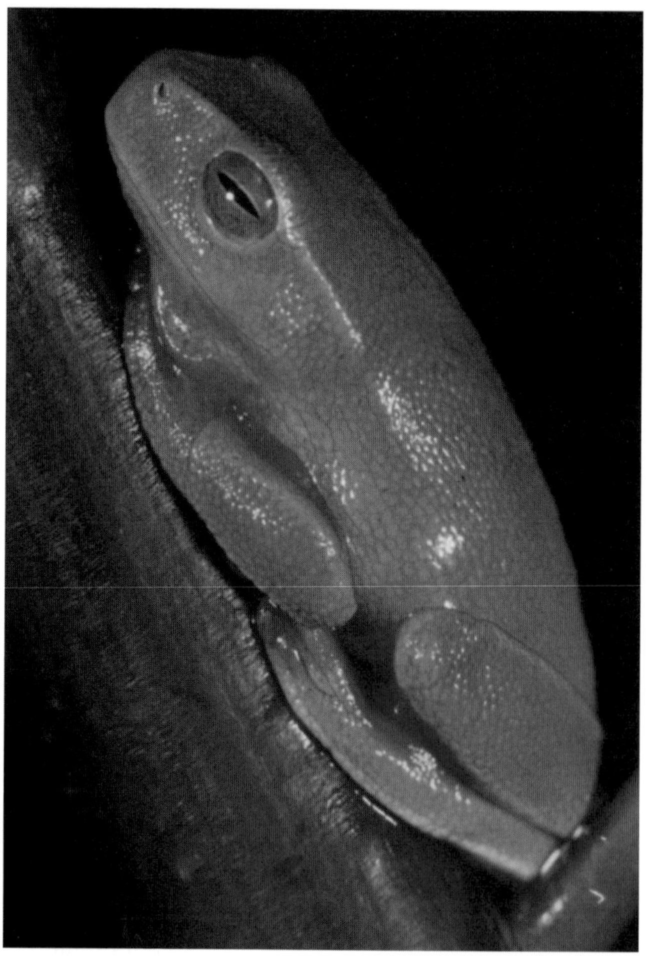

Above: Plate 33 'Acid' frogs (including Litona olongburensis) are special to Stradbroke and the other wallum areas. They are uniquely adapted to breed in waters of low pH. (Queensland Museum)

Opposite: Plate 34 The mouse-sized Feather-tail Glider (Acrobates pygmaeus) is so secretive that it has been seen by very few people. This tiny possum feeds on insects and nectar. (Queensland Museum)

Left: Plate 32 Graceful Tree Frog (Litona gracilenta) (Queensland Museum)

Plate 35 Lace Monitors or Goannas (Varanus varius) occur commonly in all habitats. They are carrion feeders, and can be seen throughout warmer months near any of the island rubbish dumps. Known as Gee-wah to the Stradbroke Aborigines. (Queensland Museum)

Plate 36 The Swamp Rat (Rattus lutreolus) is one of six species of native rats on the island. It is common in and near all the island's swamps. It is not seen often because of its nocturnal habits. (Queensland Museum)

Opposite: Plate 37 The narrow separations between the present Southport Spit, South Stradbroke, North Stradbroke and Moreton Islands have not always existed. These islands have all been linked to each other and to the present adjoining mainland at various times, due to changing patterns of sand deposition and erosion, and when sea levels have been low (see pp 14, 50-51, Appendix VI). (Department of Mapping and Surveying)

The Erosion 'Problem'

Sand bridges and the erosion 'problem': Except for small rock outcrops at Point Lookout, Main Beach Quarry, and Dunwich, North Stradbroke is composed entirely of sand. This has been deposited over millions of years by ocean currents. Patterns of deposition and erosion vary greatly. Even on a month-to-month or day-to-day timescale, beach shapes alter. All round the island's coast, lagoons, sand bars, and the beaches themselves, are built up and swept away regularly.

Since European written history of the island began, its shape and size have changed radically. It is certain that North and South Stradbroke were joined as a single sand mass until 1896, when the Jumpinpin breakthrough occurred following a catastrophic storm. Further, it is possible, even probable, that the separations at the present Southport Spit and Moreton Island are also relatively recent.

Welsby[105] records that some of the older Aborigines he knew remembered the opening of the South Passage between Moreton and Stradbroke being narrow enough for the tribes of the two islands to converse across the opening.

'...I have heard it from Toompani, a well known Aboriginal, at one time living at Amity Point for many years [Toompani died in 1888 and was buried at the back of Amity on the high sandy ridges], that his father had told him when a small boy that the blacks of Amity could in those days (the father's days) converse with the blacks of the south end of Moreton Island...Look at the formation of Amity and Moreton, and how alike they are. Who is there to say that at one time, and that not so many years ago, the now two islands did not join? And who, also, can say that that break at Southport always existed?...'

Plate 38 'The new channel at Jumpinpin, Stradbroke Island', from 'The Queenslander', 4.6.1898. The narrow link between the present North and South Stradbroke Islands was swept away in 1896,[105] following severe cyclonic weather. (Oxley Library)

The gap between Amity and Moreton Island is slowly widening over the years. The Moreton Island Environmental Impact Study of 1976 notes the erosion of the south end of Moreton at Reeders Point to be at the rate of 100 metres in every four to five years and predicts that 'much of the land below the 5 metre contour south and south-east of Kooringal could be eroded within 20 years'.

Cook sailed by what is now North Stradbroke in 1770. He did not refer to it as an island, recording neither The Spit break (between what is presently South Stradbroke and the mainland near Southport), nor the South Passage Bar break between North Stradbroke and Moreton (which was first reported by Flinders in 1799; see Appendix VI). In the light of this information, with the knowledge that it was possible to call between Moreton Island and the present Amity area in fairly recent times, that a sea level drop of only 28 metres would re-link the areas by moving the present coastline east of both Moreton and Stradbroke, it is reasonable to suggest that Moreton, North Stradbroke, South Stradbroke and the Southport Spit were all joined or closely connected in fairly recent times and in geological times when sea levels were low. 'Moreton Bay' at that time was totally protected along its eastern boundary by a long, narrow strip of sand. The affinities of the fauna and flora for each area confirm this suggestion.

Erosion of the island's foreshores has always occurred. The Aboriginal inhabitants of the island treated this as a natural and accepted phenomenon. They accounted for South Passage break-through and the erosion of Amity's shores in legends that have been preserved and made available through the care and diligence of the late Mr A Gunn of Wynnum.

Plate 39 Foreshores at Adder Rock c1980, following cyclonic weather. All round the island, beach profiles change continuously as sand accumulates and then is swept away again. (Authors)

The legend of South Passage: Many years ago when the islands of Minjerriba [Stradbroke] and Moorgumpin [Moreton] were one, there lived an old man named Mirapool, the wind caller.

He held a shell which was passed down from generation to generation of wind callers and to which was attributed his power over the winds.

The people of Moorgumpin became so jealous that they sent two men to steal the shell, as they supposed after the loss of his shell his power over the winds would cease.

When Mirapool discovered his loss he became so enraged that he decided to enlist the aid of the winds to create a passage of water over which the people of Moorgumpin would never pass.

Standing on the dividing line, he called first to the North wind, then to the South wind, then to the East wind, then to the West wind; and as many nights as can be counted on one hand the wind raged and fought, with Mirapool screaming harder and harder.

His people became afraid, and on the third day implored him to stop. He heeded them not, and each day returned to his place on the hill calling for the winds to blow stronger and stronger.

The winds died down and the Nooghies and the Noonukuls found that they were separated by a sheet of water which did not exist before the battle of the winds and this division between the islands where the ocean meets still water is now known as the South Passage.

The legend of Amity's foreshore: No man, not even Mutallo the old one, who had twice seen the fire in the sky [Halley's Comet] could remember when the Nooghies of Gnoorganpin commenced to call to the Noonukuls that they would come to drive them into the hills. But the Noonukuls were strong and feared them not. Daily they hurled insults back across the narrow waters of South Passage. Then to the tribe of Compulla, the fish caller of the Noonukuls, came a strange and lasting sickness which laid many dead and many low. The Nooghies, hearing no response to their calls, sent Panderook the strong to see why no answering call came back. Looking down in the light of a half moon he saw the once great tribe of Compulla lying scattered around the fires. Hastening back to Gnoorganpin he told of what he had seen. With no delay the Nooghies crossed the narrow waters, and as they came they cried 'We have come to drive you back into the hills.' From one fire rose Wanderoo the fire maker, and facing the Nooghies he cried 'My people have lived here since the very dawn of time, and you are the first to come not in peace. To you or any other man this island will not belong, for day by day, month by month, year by year, this island will disappear into the sea until only the rocky headlands of Mooloomba remain.'[83a]

European settlers, landholders, and governments have been plagued by the erosion 'problem' at Amity. Much of what was Amity township is now far out in the Rainbow Channel. At least 350 metres of sand on which Amity was initially established were swept away between the 1820s and 1950s. Early surveyors recognised the 'problem', which is in fact a natural phenomenon, from the start. They coped by surveying long, narrow blocks so that buildings could be moved back periodically from the encroaching sea.

Plate 40 The sea takes back some of its own at Amity Point, c1980. (Authors)

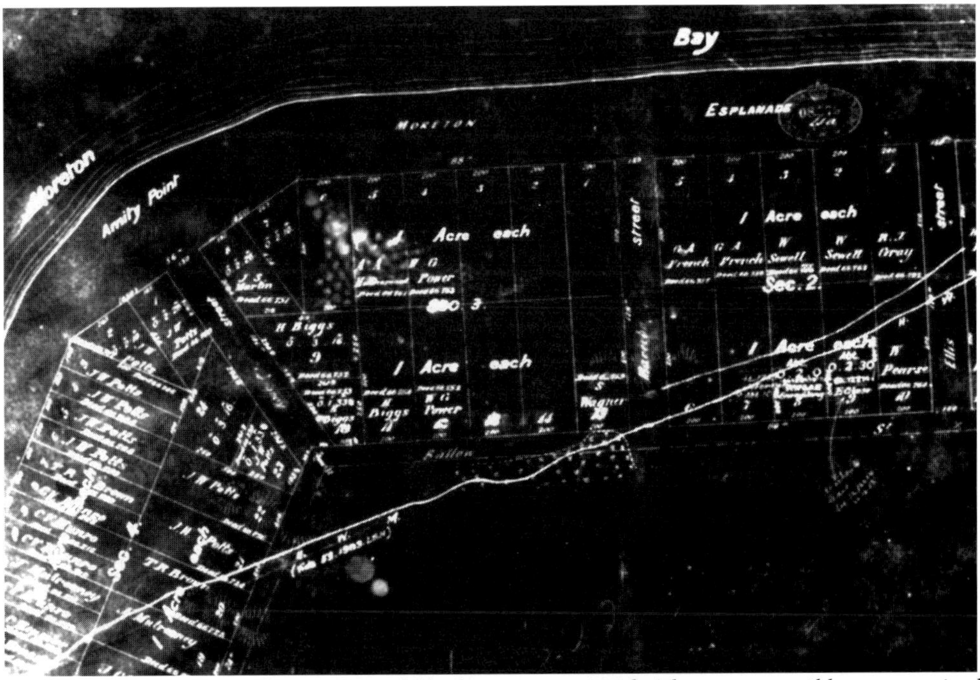

Plate 41 The first survey for residential blocks at Amity (1886). The surveyor, Abbot, recognised Amity's erosion 'problem'. He made provision to shift dwellings back from the encroaching sea by ensuring blocks were deep. See fig 21. (State Archives, A339-1).

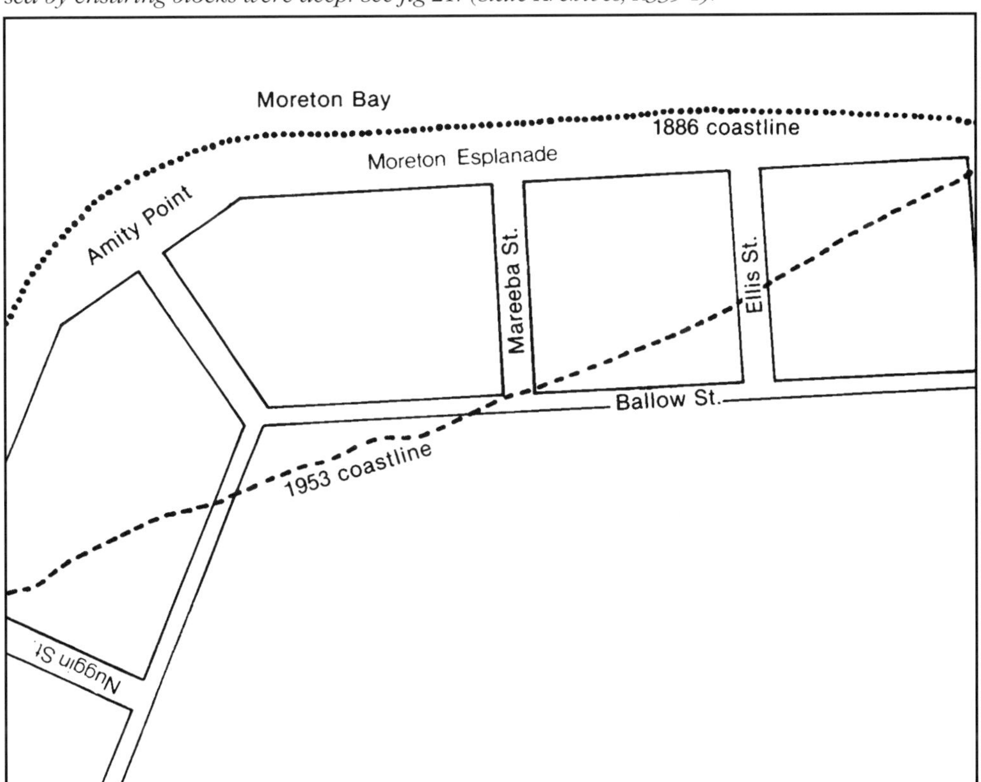

Fig 21 The first survey of Amity, 1886. See plate 41.

The Social Environment

Plate 42 With the destruction of Aboriginal midden complexes by mineral sandmining operations, irreplaceable records of a social history – important to both Aborigines and non-Aborigines – has been lost. (Authors)

Thousands of years before Lieutenant James Cook 'discovered' Stradbroke on 17th May, 1770, Aborigines lived successfully on the island, utilising but not devastating its natural resources. Initial contacts between the two races were cordial. Matthew Flinders recorded in 1803 'twenty Indians upon the side of a hill near the shore, who seem to be peaceably disposed, amusing us with dances in imitation of the Kangaroo...We made signs of wanting water which they understood, and pointed to a small rill falling into the sea.'[40] Four men landed in the vicinity of Cylinder Beach to replenish the vessel's water supplies.

There was no more contact for 20 years, when timber-getters Pamphlet, Parsons, and Finnegan were blown off-course and wrecked on Moreton Island. They were transported to Stradbroke by friendly Aborigines from Amity and were the first white men to live with the Aborigines in what is now Queensland. The castaways recounted that 'They placed us in a very large well-built hut by ourselves, and supplied us with fish, water etc. very liberally. Here we remained for a week or ten days, during which time we were most hospitably treated by the natives. They would not, however, suffer us to approach the huts in which their women were, for the first five or six days; and at night five or six of the younger men would sleep in front of the hut. But they afterwards became less vigilant, and we used to pass through their huts among the women as we pleased.'[64]

Subsequent early contacts were friendly and apparently based on mutual respect. '...[T]hey are a quiet, inoffensive, good natured, lively set of people', Oxley notes. 'The natives...appeared to possess a most friendly disposition'.[105] And in 1847, when the 'Sovereign' was wrecked on South Passage Bar, Aborigines gallantly assisted survivors ashore.

The Noonukul were the traditional owners of the northern part of Stradbroke; but since first European contact at least, the people who lived on the island came from many areas of southeastern Queensland and mixed socially with people from at least as far away as the Bunya Mountains. Numbers were never high. The early estimates of the maximum number of Stradbroke people is 300.

In 1827, Captain Logan established a convict and military depot at Dunwich. Four Passionist priests were sent in 1843 from Europe to set up a mission to convert the Aborigines. With the closure of the Dunwich depot, the vacated buildings, which were in a state of disrepair,

Fig 22 North Stradbroke Island, localities mentioned in the text

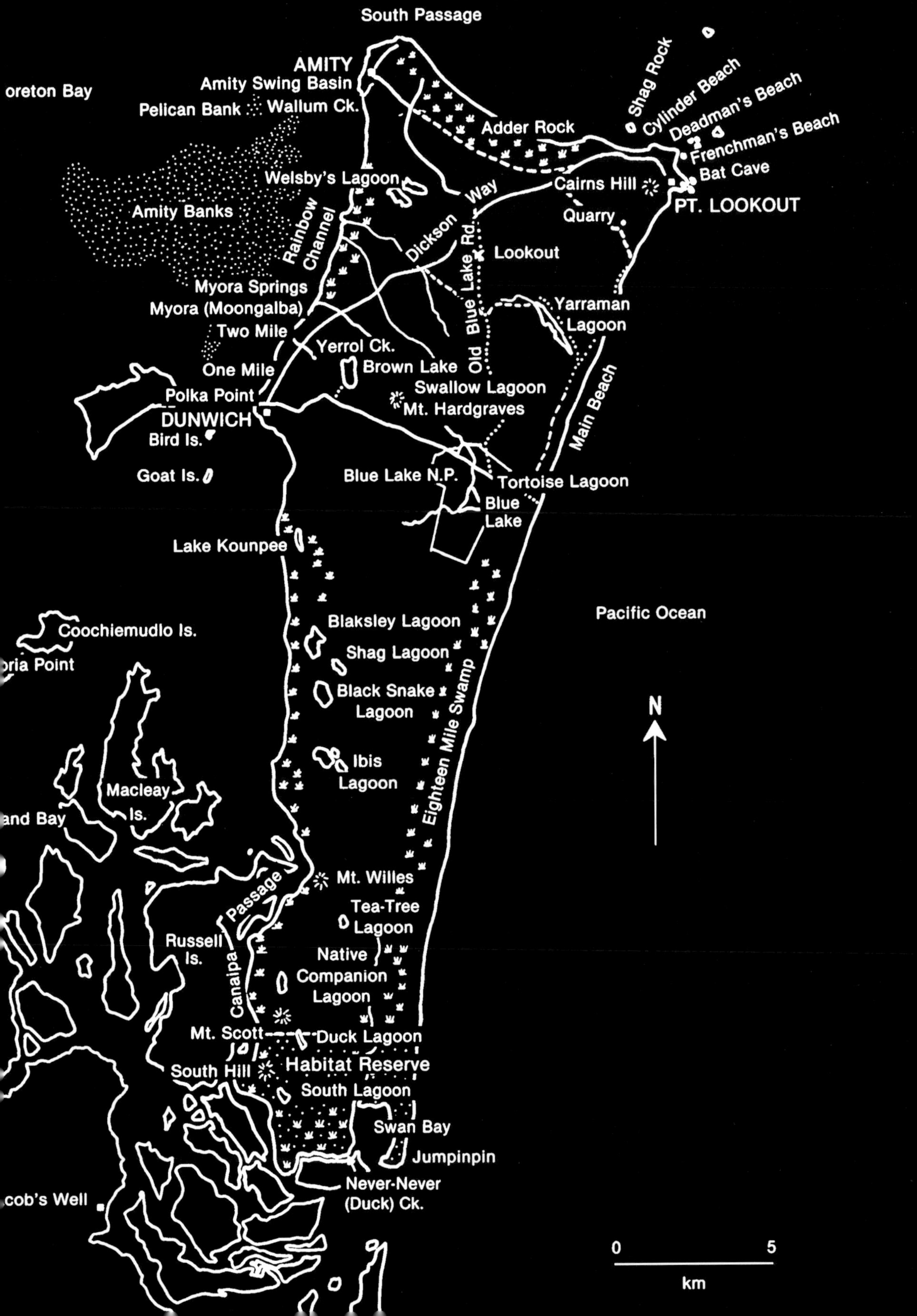
South Passage
AMITY
Amity Swing Basin
oreton Bay
Pelican Bank
Wallum Ck.
Shag Rock
Cylinder Beach
Deadman's Beach
Frenchman's Beach
Adder Rock
Bat Cave
Cairns Hill
PT. LOOKOUT
Welsby's Lagoon
Amity Banks
Rainbow Channel
Dickson Way
Quarry
Lookout
Old Blue Lake Rd.
Myora Springs
Myora (Moongalba)
Two Mile
Yerrol Ck.
Yarraman Lagoon
Brown Lake
One Mile
Swallow Lagoon
Polka Point
DUNWICH
Mt. Hardgraves
Main Beach
Bird Is.
Goat Is.
Blue Lake N.P.
Tortoise Lagoon
Blue Lake
Lake Kounpee
Pacific Ocean
Blaksley Lagoon
Coochiemudlo Is.
oria Point
Shag Lagoon
Black Snake Lagoon
Eighteen Mile Swamp
N
Ibis Lagoon
Macleay Is.
and Bay
Mt. Willes
Passage
Tea-Tree Lagoon
Russell Is.
Canaipa
Native Companion Lagoon
Mt. Scott
Duck Lagoon
South Hill
Habitat Reserve
South Lagoon
Swan Bay
Jumpinpin
Never-Never (Duck) Ck.
cob's Well
0
5
km

were taken over by the priests. The Mission was considered a failure, and in 1847 the priests were recalled. Nine desperate convicts, who had escaped in 1853 from Norfolk Island, landed at Amity from where they stole the stores of Fernandez Gonzales, a dugong fisherman. The convicts caused havoc in Moreton Bay until they were recaptured and returned to prison 16 days later. One of the colourful Europeans that interacted with the Aborigines was Gustavus Birch who lived at Amity from 1850-1880. He recorded Amity life in his diary.

Missionaries, colonial attitudes, and intermarriage quickly wreaked havoc on tribal life. Two unsuccessful attempts by the government to establish an Aboriginal mission on Bribie Island failed for a number of reasons, one being that the sites were unsuitable. In 1892, the Bribie Mission was moved to Myora-Moongalba, which was a traditional site. The Mission officially operated until 1942.

The first permanent European residents of Stradbroke came to Amity in the 1820s, probably about 1825, to establish a pilot station to guide ships to Brisbane through the South Passage Bar, between Stradbroke and Moreton Islands. There are presently about 3000 permanent residents on the island, with possibly 40,000 tourists swelling the population in peak holiday seasons.

Plate 43 Janie Sunflower, born at Polka Point, Dunwich (Goompie), was given the native name Wynyeeba. While quite young, she was sent to the Mission on Bribie Island, but was later moved back to Stradbroke where she attended the Moongalba Mission School. Here she lived until going to Cherbourg Settlement in 1910. After marrying at Cherbourg, she and her husband worked on several stations in southwestern Queensland, until her husband's death, when she returned to Cherbourg to teach the kindergarten classes. After ten years at the settlement, she returned to Dunwich in 1957 and lived there until her death in 1964. She was buried in the Dunwich Cemetery. (Helen Elmslie)

Plate 44 Midden (or Aboriginal rubbish dump) on Main Beach before sandmining. There is nothing like this left on the eastern foredunes today. Further, most of the western dune middens have been destroyed or modified by sandmining operations. In the 1960s, extensive sandmining destroyed hundreds of shell middens of cultural significance. (Doug Hirst)

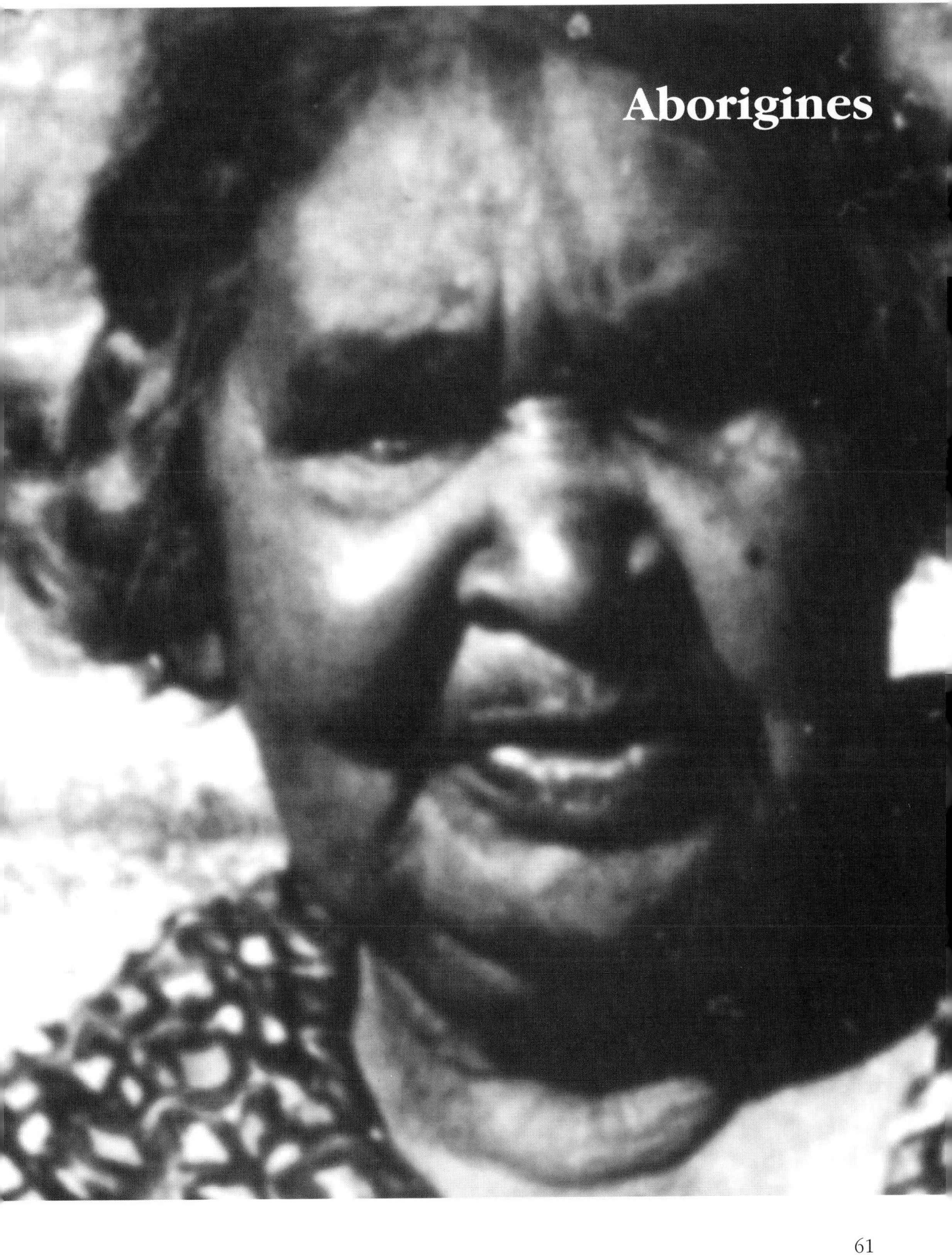

Aborigines

The original owners of Stradbroke lived well on the island. Its plants provided means of transport, shelter, and a reliable food source; its seas teemed with fish, shellfish, and dugong; and the island supported a rich vertebrate fauna which could provide additional food. Here, as in the rest of the Moreton Region, in a richly endowed, mild environment, the Aborigines were semi-sedentary. In many other parts of Australia, poorer conditions made nomadic life a necessity.

Shelter: The Stradbroke Islanders appear to have built two types of shelters. One, about 3 metres across and 1.2 metres high, held about nine people and was a semi-permanent dwelling. With a fire inside, the huts were warm in winter. A series of four evenly-spaced crossing hoops of wattle or other pliable saplings was firmly embedded in the ground. The spaces in each segment, with the exception of one that served as a door, were filled with extra pieces of wood which were tied to the hoops. The whole was made waterproof by overlapping sheets of Tea-tree bark kept in place by heavy sticks laid against the outside. This type of shelter seems to have been restricted to coastal southeastern Queensland.

A smaller, less permanent shelter was made by cracking one sapling which was supported in a forked stick. Other sticks were added, forming a peak. Covering Tea-tree bark was laid longitudinally. The bark was weighted with heavy sticks.

Plate 45 One of the traditional shelters constructed by Aborigines in coastal southeastern Queensland. This semi-permanent structure was photographed on Bribie Island, in 1894. (Oxley Library)

Plate 46 Native huts, Moreton Island. Painting by Owen Stanley, 1847 (Mitchell Library)

Plate 47 Tow-row fishing, Burpengary Creek, in the 1890s. Tow-row nets were a 'deep long pocket attached to a frame in the shape of a bow...a net was held in each hand with the string side of the bow towards the body and the pocket trailing behind'[112]
'...The method of handling the nets brought the handle across the top of the wrist of each hand, causing a callous from constant rubbing'.[21] (Oxley Library)

Plates 48 Shellfish were a major component of the diet of coastal Aborigines. They collected eugaries on ocean beaches, and whelks and oysters in the mangroves. This photograph was taken in 1912, in mideastern New South Wales. It shows typical food gathering activities for tribes in coastal Queensland and New South Wales. (Queensland Museum)

Food: The sea provided the bulk of the islanders' food. Middens (Aboriginal rubbish dumps), consisting of multiple layers of discarded shell from a variety of molluscs, are evidence of the numbers of shellfish eaten. The eugarie is the chief shell of the ocean beach middens, while those of the western shore are made up of a mixture of oysters, cockles, and periwinkles. There were many of these dumps and, in places, they were several metres deep. Ponosov[82] recorded 121 sites on North Stradbroke and 28 sites on South Stradbroke. One large remaining midden at Polka Point, Dunwich, is exposed along the eroded beach face. Most middens have been destroyed by sandmining operations.

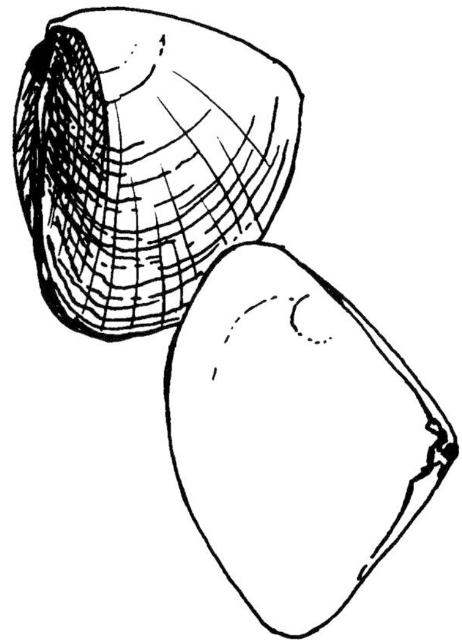

Fig 23 Eugarie (Donax deltoides)

Petrie tells of the great seasonal catches of mullet. 'Mullet, an important sea food, was usually caught in tow-rows, which were pairs of nets somewhat heart-shaped and about 1.2 metres across. A group of men would combine to make a wall of net in front of a sighted school of fish, which would be then driven towards them by other Aborigines in canoes.'[80]

Fish surplus to immediate needs was wrapped in soft grass twisted into a fine rope and wound closely around the fish to exclude flies. The fish, stored in dillies hung in trees, kept well. Petrie claimed that during the cold weather he had eaten fish two weeks old.

To prepare the fish, the women scaled them with a eugarie shell before placing them on a bed of hot coals. When partly cooked, the shrunken and hardened gut was pulled out with a finger through the gill opening.

A remarkable relationship existed between Aborigines and dolphins. The Aborigines did not kill them. Petrie relates how the Aborigines enlisted the aid of dolphins to catch tailor, 'punta', when they travelled in shoals like the sea mullet.

'The sea would be calm and there would be no sign anywhere of a porpoise ('talobilla'), the blacks would go along the beach jobbing with their spears into the sand under the water, making a queer noise, also beating the water with the spears. By-and-by, as if in response, porpoises would be seen as they rose to the surface making for the shore and in front of them schools of tailor fish. It may seem wonderful, but they were apparently driving the fish towards the land. When they came near, the blacks would run out into the surf, and with their spears would jab down here and there at the fish, at times even getting two on one spear, so plentiful were they. As each fish was speared, it was thrown to shore, and there picked up by the gins. The porpoise would actually be swimming in and out amongst all this, apparently unafraid of the darkies. Indeed, they seemed rather to be all on good terms, and I have with my own eyes, more than once seen a black fellow hold out a fish on a spear to a porpoise, and the creature take and eat it.'[80]

(The term 'porpoise' was loosely used. Porpoises are not found in Australian waters.)

Mr John Campbell is quoted by Petrie's daughter. 'Doubtless this statement about the porpoises and blacks fishing together will be pronounced – as I myself did upon hearing it – to be a myth; in fact all nonsense; but further inquiry and observation has convinced me that it was a fact, and any persons doubting it can convince themselves by going to Amity Point during the fishing season. The blacks even pretended to own particular porpoises, and nothing will offend them more than to attempt to injure one of their porpoises.'[80]

H J Hall confirms the veracity of these reports and writes of similar instances of dolphins aiding humans fishing in Africa, South America and Asia.[44a]

The dugong was sought for its flesh. Europeans found the flesh equally attractive, and hunted it almost to extinction. While the Moreton Bay population has slowly increased and is, once again, secure, boat propellers are responsible for dugong and turtle deaths. Dugong feed on marine grasses which grow on

Plate 49 North Stradbroke Island Aboriginal stone artefacts owned by Miss A Tripcony. (Authors). As rock occurring on Stradbroke was not suitable for making stone implements, the Aborigines traded source rock and implements with their neighbours on Moreton Island and the mainland.
a – sandstone mortar and pestle used for grinding and pounding vegetable foods
b – sandstone mortar
c – flaked axe of volcanic material. Axes like this were hafted on wooden handles and used as tools for many purposes, including stripping tree bark to make canoes and huts.

the tidal shallows of Moreton Bay. Shreds of floating sea grass were an indication to the Aborigines of dugong.

Petrie has described the dugong hunt. A couple of Aborigines in a canoe drove the dugong towards a strong large-meshed net held by a number of men who rolled the trapped animal round and round in the submerged nets until it drowned The dugong net was made from the native cotton tree's inner bark rolled into a strong twine. The prepared cord was meshed into a net with a large net needle. Matthew Flinders found one of these large nets, '...the meshes of which were wide enough to admit the escape of a moderate size porpoise; and the line of which it was made was from three-quarter to an inch in circumference.'[41] He collected the net in 1799 near what is now Redcliffe while he was preparing charts of Moreton Bay.

The Quandamooka Elders tell how the capture of a dugong always meant a great feast. The women were not allowed to look upon the captured animal before it was cut up. Today, with protective legislation, on Stradbroke, only traditional owners can hunt dugong and turtle for traditional use. The Quandamooka Dugong and Turtle Group make decisions on the number that may be taken.

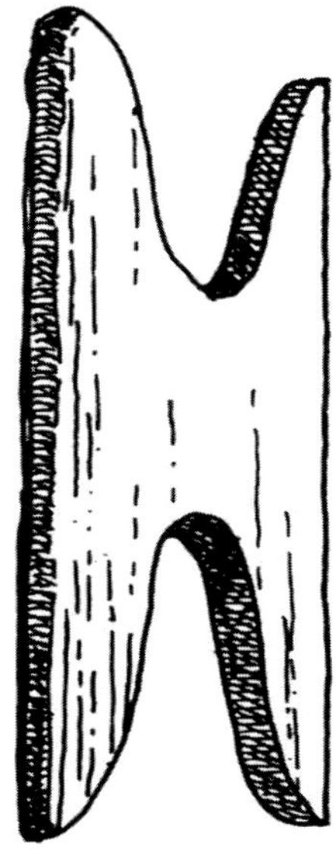

Fig 24 A 'needle' used in making dugong nets

Plate 50 Bob Campbell and a Dugong catch, Amity, 1940s (Gregory family)

Grey kangaroos, wallabies, bandicoot, echidna, goannas, flying foxes, swans, parrots, and other birds were hunted and eaten. The Carpet Snake (ka-bool), the totem of the Noonukul people, was taboo for eating. (Mrs Janie Sunflower, personal communication)

The toasted or raw larvae of a large moth (*Xyleutes* sp) was a delicacy. The grub burrows into tree trunks. It was hooked out of its hole with some suitable twig. The larvae were located by their tell-tale deposit of sawdust-like excreta. These succulent grubs, generally known as witchettys, on Stradbroke are called 'jubbums'.

The rhizomes of the ferns (*Blechnum* spp), which are common in the freshwater swamps, were collected and prepared in great quantities by the women. Backhouse, a Quaker minister visiting the colonies to teach the gospel, was detained in 1836 at the Amity Point Pilot Station while he waited for calm weather for his ship to cross the South Passage Bar. He was a keen naturalist, and described the preparation of the fern root which the Aborigines called 'bungwal' or 'dingowa'.

'This, after it was roasted, was held in one hand on a log of wood, while its whole length was beaten, by a stone, held in the other hand, so as to break the woody fibre. In this state it is eaten, without removing the charred surface; its taste is something like a waxy potato, but much more gelatinous.'[3]

The rhizome was also stripped of its outer coating, dried in a slow fire, then pounded between two bungwal stones into a flour which was made into something resembling a damper.

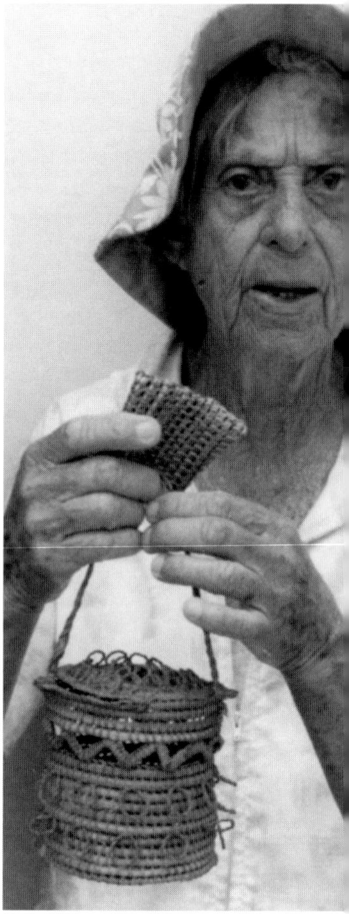

Plate 51 Martin basket held by its owner, Miss Anastasia Tripcony of Dunwich. (Authors)

Berries were collected as food. The Midyim (*Austromyrtus dulcis*), a small low-growing shrub producing small attractive white-grey berries with mauve markings, was eagerly sought. Backhouse records his impression of the berry as being 'the most agreeable, native fruit I have tasted in Australia; they are produced so abundantly, as to afford an important article of food to the Aborigines'.[3]

The ripe fruit of the Screw Pine or 'wynnum' (*Pandanus* sp) was chewed at the soft end. The greater part of the fruit is hard and woody.

Weapons and implements: Boomerangs, spears, nullas and shields were used. The boomerangs were returning and non-returning. Most of the stones used for implements were traded with mainland tribes. Axes were ground from mainland igneous rocks. The grass-tree gum, a form of shellac, was used to fasten axes to handles.

The water-worn stones found in the inlets below Cape Moreton were sought by the Stradbroke Islanders for preparing the bungwal (*Blechnum indicum*).[21]

Baskets: Baskets were woven from reeds by women The reeds ('moon-ga') were first stripped, soaked in running water, dried, and processed in hot ashes. Backhouse describes the process. 'The base of these rushes is of a pale colour, the portion included in the sheaths at the base, or just emerging from them, is of a pinky hue and the top green. By arranging the knots so as to form diagonal lines across the bag, the colours are brought into tasteful order.'[3]

These baskets were called 'kulai' on Stradbroke. On Bribie they were known as 'dilli'. The skilfully woven baskets made by the coastal women were sought after by inland traders.

Plate 52 Basket made by Mrs Martin (Nunninugh). Such artistic reed baskets were made by coastal Aboriginal women throughout southeastern Queensland. (Authors)

Canoes: For canoe making in southeastern Queensland, the bark of the 'bastard mahogany' was preferred, according to Petrie, because it was less prone to splitting when dry. Whether 'bastard mahogany' is Swamp Mahogany (*Eucalyptus robusta*) or the Box (*Tristania*) cannot be established with any certainty. Both grew in coastal swamps. The bark was carefully prised off the tree in one large piece. Vines were tied either end to ensure the bark would retain its shape. The hollow was filled with sticks which were lit. The heat made the bark more pliable and easier to shape. The ends were retied and the crevices filled with clay. The large canoes that carried nine to ten men had supportive wattle frames.

Plate 53 The grave of Sydney Rollands, mother of Sam Rollands, who was one of the last of the island Dugong fishermen. This lies on the hill above the site of the Moongalba Mission. (Authors)

Missions: Bishop Folding's concern for the spiritual wellbeing of the Aborigines resulted in the establishment at Dunwich in 1843 of the first recorded Catholic Mission to the Aborigines. The four Passionist Order priests were recruited from Europe. The mission was set up in the dilapidated deserted buildings of the depot set up in 1827 by Captain Logan near the present Dunwich jetty and barge loading complex. From the beginning, problems beset the new mission which was run by Father Raymond, a poor administrator. The missionaries had difficulty communicating with the Aborigines. Only one of the priests could speak English. The others were poor linguists. Following seasonal foods, the Aborigines frequently moved away from the mission site. The mission's isolation made his task all the more difficult. When the mission failed and the priests were directed elsewhere, Father Raymond refused to go. He remained another year. The 'Moreton Bay Courier' of April 24th 1847 reported that Father Raymond, no longer in a position to supply the Aborigines with food, was threatened by them. He appealed to the authorities for protection.[104]

W M Tyson, Secretary of the Queensland Aboriginal Protection Association and Superintendent of the Bribie Island Mission, reports the disbanding of the Bribie Mission in 1892. The unsuitably sited mission was moved to Myora, also known as Moongalba, Stradbroke Island. As there were no buildings on the proposed new site, the SS 'Miner' transported the Bribie people along with their belongings to Peel Island, until buildings were constructed at Moongalba. Many Aborigines from southern Queensland settled at Moongalba.

Material culture items today: Very few of the artefacts made by the Aborigines have been preserved. What remains, including middens, is now totally

Plate 54 Early days, Moongalba Mission (Oxley Library)

Plate 55 Children at the Moongalba Mission, 1896 (Oxley Library)

protected under the Aboriginal Relics Preservation Act. All material remains of Aboriginal culture are the property of the Crown, apart from implements collected within Queensland or interstate before 1967.

The aim of the legislation is to ensure the preservation and protection of the State's cultural heritage for future generations of both Aborigines and other Australians.

The Cultural Heritage Unit, Department of Natural Resources and Mines controls the identification and protection of Aboriginal sites.

Aboriginal names in use on Stradbroke today: Many of the place names in use on Stradbroke today are those used by the Aborigines of the island. Some are set out below.

Currigee	South Stradbroke
Minjerriba	North Stradbroke Island
Yarral	Creek at Two Mile
Capeembah	Creek at Myora
Goompie	Dunwich
Pul-an	Amity
Dumba	Stradbroke Island
Bummel	Brown Lake
Karboora (Gubberr)	Blue Lake
Moonpanbilla	Bird Island
Kindara (Gindava)	Hills of Moreton
Mooloomba	North end of Stradbroke (Point Lookout)
Booran	South
Timbin	North wind
Billa	Swamp Oak (*Casuarina glauca*)
Mintee	*Banksia integrifolia*
Bamberra (Bambara)	*Banksia serrata*
Midyim	*Austromyrtus dulcis*
Yarrong	Sand
Garlen (Galeen)	Cloud
Bigoon	Jew fish

Plants used by the Aborigines: Plant roots and tubers used on Stradbroke were roasted in hot coals. Prior to European contact, the Aborigines did not know how to boil water. When the castaways Parsons, Finnegan and Pamphlet emptied a pannikan of boiling water on the ground, the Aborigines were frightened.[64a] Poisonous seeds and roots were made edible by soaking in water. A netted dilly was used to hold the seeds and roots during soaking.

An asterisk denotes that the information relates specifically to Stradbroke. Stradbroke Aboriginal names are given in brackets following the common name where these are available.

> Pigface (*Carpobrotus glaucescens*): edible fruit and leaves
> Grey or White Mangrove* 'tehuntehee' (*Avicennia marina* var *australasica*): edible fruit, shields
> Coast Morning Glory (*Ipomoea cairica*): edible tubers
> Velvet Beans (*Mucuna gigantea*): edible seeds
> Sand Paper Fig (*Ficus coronata*): edible fruit; rough leaves used for final smoothing of weapons
> Small-leaf Moreton Bay Fig* 'kunnen' (*Ficus obliqua* var *petiolaris*): edible fruit
> Midyim* (*Austromyrtus dulcis*): edible fruit
> Lilly-Pilly (*Acmena smithii*; *Syzygium coolminiana*): edible fruits
> Yellow Wood Sorrel or Sourgrass (*Oxalis corniculata*): edible leaves
> Bangalow or Picabeen Palm (*Archontophoenix cunninghamiana*): young shoots edible; flower sheaths used as carrying vessels
> Cabbage Tree Palm (*Livistona australis*): top shoots pulled out and eaten raw
> Paper Bark* 'gnoojoorpa' (*Melaleuca quinquenervia*): bark used for waterproofing huts
> *Podocarpus spinulosus*: edible fruit
> Screw Palm* 'wynnum' (*Pandanus pedunculatus*): edible fruit and young leaves. The ripe fruit…'turns orange-yellow to orange-red. A small amount of sweet pulp can be obtained by chewing the fibrous semi-fleshy lower part of the ripe segment. However, depending on the species, mild to severe irritation to mouth and throat may follow. The Aborigines destroyed the irritating principle by roasting the fruits before they chewed off the flesh or scraped it into water to make a sweet drink…Beneath the fibrous layer the fruit is hard and woody, with several longitudinal cavities each containing a single oily seed. The seeds make pleasant eating, but one requires a tomahawk to extract them. The Aborigines roasted the fruit before extracting the seed…'[28]
> Tree Orchid (*Cymbidium madidum*): pseudobulbs chewed to combat dysentery
> Ladies' Tresses (*Spiranthes* sp): tubers considered a delicacy; tubers of all ground orchids dug and eaten
> Wallum Banksia or Dungle Nut* (*Banksia aemula*);
> Coast Banksia* 'tchoomeroo' (*Banksia integrifolia*);
> Dwarf Banksia* (*Banksia oblongifolia*);
> Red Honeysuckle* (*Banksia serrata*):
all have a honey-like nectar which may have been soaked in water to make a sweet drink
Geebung* 'koom-barra' (*Persooma cornifolia*; *P media*;* *P virgata*;* *P linearis**): edible fruits which were eaten raw. The Aborigines 'swallowed the pulp and stone which they squeezed from the skin with their fingers.'[80]
> Small-leaf Raspberry (*Rubus parvifolius*): edible fruit
> Five Corners* 'gutta-rulla' (*Styphelia vindis*): edible fruit
> *Timonius timon*: edible fruit

> *Simlax glycyphylla*: tonic, antiscorbutic
> Bribie Cypress* 'brorogorre' (*Callitris columellaris*): resin; roots used for boomerangs
> *Callitris rhomboidea*: resin
> Cotton Tree* (*Hibiscus tiliaceus*): inner bark twisted into net twine
> Cunjevoi (*Alocasia macrorrhiza*): roots eaten:
'The plant is poisonous, but the Aborigines prepared the roots by soaking them a long time, and then they were pounded up and made into cakes, and so roasted on the cinders.'[80]
> Asthma Plant (*Euphorbia hirta*): remedy for asthma and bronchitis
> Milky Mangrove (*Excoecaria agallocha*): sap a cure for ulcers and chronic diseases
> Wattle (*Acacia sophorae*): seeds edible
> Tape Vine (*Stephania japonica* var *discolor*): fish poison
> Mistletoe* (*Amyema* sp): sweet sticky edible berries
> Mistletoe* (*Lysiana exocarpia*): chewing-gum made from the sweet sticky berry. 'When quite ripe, the sticky, gelatinous, sweet pulp surrounding the single seed is pleasant to eat but its great sticky mess often makes the seed almost impossible to spit out.'[28]
> Native Cherry 'tchimmin-dillen' (*Exocarpos cupressiformis*): fleshy fruit stalk eaten
> Bulrush (*Typha muelleri*): young shoots and root edible.
'The outer skin was taken off, and then the roots were chewed until nothing was left but fibre which was thrown away.'[80]
> Grass Tree* 'dak-ka-bin' (*Xanthorrhoea arboea*; *X* sp aff *arborea**; *X australis**; *X johnsonii**; *X macronema**; *X media**; *X pumilio**; *X resinosa fulva**): base of inner leaves edible, resin
> Matrush (*Lomandra longifolia*): fibre
> Native Ginger (*Alpinia caerulea*): edible fruit
>Bracken Fern (*Pteridium esculentum*): edible rhizome
>Dingowa or Bungwal Fern* (*Blechnum indicum*): edible rhizomes, an important component of the diet on Stradbroke
> Native Lasiandra* (*Melastoma polyanthum*): edible fruit that blackens the tongue
> Supplejack* 'yurol' or 'yeroll' (*Flagellaria indica*):
'The stems were used by the natives in tree-climbing. They peeled off the rind, after which the two ends of the vine could be knotted together to form a kind of hoop which they used to climb the trunks of tall trees and to sit in when chopping out the nests of bees.'[4]
> Pigweed (*Portulaca oleracea*): edible; antiscorbutic; seeds also eaten
> Crabs' Eyes* 'gidgee' (*Abrus precatorius*): probably introduced to the island by the Aborigines who used seeds for ornamentation. The seeds are poisonous.
> Freshwater Reed* (*Baumea rubiginosa*): basket making
> Dogwood (*Jacksonia scorparia*): gum eaten
> Red Ash* (*Alphitonia excelsa*): leaves used for poisoning fish; possibly 'Pigeon Berry' mentioned by Welsby.[104]

European History

Previous page: Plate 56
Captain James Cook

Discovery: Lieutenant James Cook records in his journal of 17th May, 1770: 'Between 4 and 5 we discovered breakers on our larboard bow, our depth of water… was 37 fathom. At sun set the northermost land in sight bore NBW, the breakers NWBW distant 4 Miles, and the northermost land set at noon, which form'd a point which I named Point Look-out, bore west distant 5 or 6 Miles (Latitude 27° 6'). On the north side of this Point, the shore forms a wide Bay which I named Moreton's Bay, in the bottom of which the land is so low that I could but just see it from the top mast head. The breakers I have mentioned lies about 3 or 4 Miles from Point Look-out. At this time we had a great sea from the southward which broke prodigious high upon them…'[9]

In 1803, Matthew Flinders set sail for England as a passenger on the 'Cato'. The ship was wrecked on the central Queensland coast at Wreck Reef, east of where St Lawrence is today. In the ship's cutter, which the men christened 'Hope', Flinders returned to Sydney for help. With him he took 12 men, six to row in two shifts, and the captain of the 'Cato'. He reached Sydney in two weeks. On his trip south, he sent four men ashore at Cylinder Beach to replenish his water casks. Flinders wrote that he then attempted to continue his journey south past Point Lookout. A southerly wind forced him to anchor for the night: '…under the lee of two rocks which lie 1 1/2 mile to the N.W. ward from Pt. Lookout: the surf prevented us from landing on the rocks, otherwise many birds might have been caught in the night, but we caught fish enough to make a good meal.'[40]

The rocks were the Shag Group. At daylight Flinders continued south past Point Lookout.

Neither Cook nor Flinders described Stradbroke as an island. This was done first by John Oxley. He noted how Moreton Bay extended south to open into the sea at what is now Southport, in 1823.

White men on Stradbroke: The castaway timber-getters, Pamphlet, Parsons, and Finnegan, wrecked on Moreton Island in 1823, crossed to Amity Point in an Aboriginal bark canoe. These were the first known white men to have lived with the Aborigines in what is now Queensland. From a tree at Myora, they constructed a dugout canoe using a hatchet salvaged from their boat, and, after spending approximately six weeks on Stradbroke, launched their canoe for the mainland. Believing they were south of Sydney, the castaways walked north – away from their goal. John Oxley, searching for a suitable site for a new penal settlement, rescued Pamphlet and Finnegan from Bribie Island.

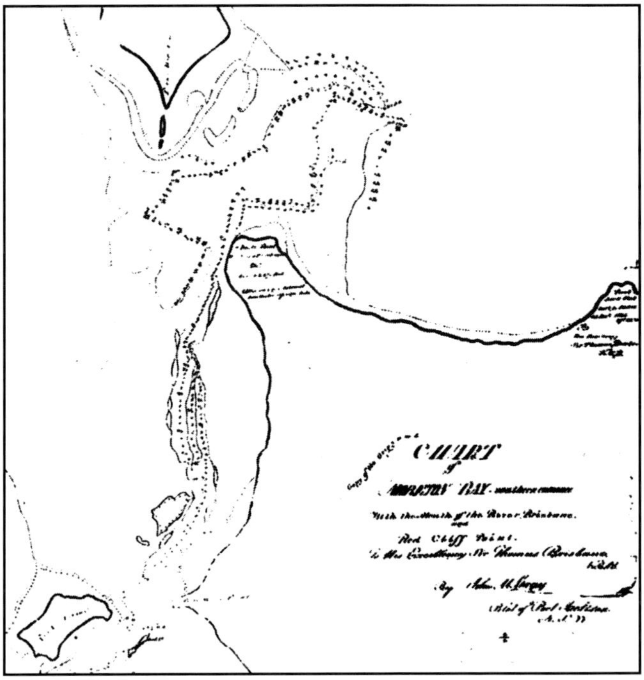

Fig 25 The first survey of South Passage Bar, by John Grey, 1824 (State Archives 1A1)

The men told him of a large river flowing into Moreton Bay. Parsons at the time was absent on a trip north with some of the Bribie tribe. Oxley left a note in a bottle for the absent Parsons. The note assured Parsons, who could not read, that Oxley would be back to rescue him. Oxley, accompanied by Governor Brisbane who wished to inspect the newly-found Brisbane River, returned to collect Parsons the next year in the government brig 'Amity'. They also visited Peel Island and Point Lookout. Oxley and surveyor Robert Hoddle surveyed the South Passage between Moreton and Stradbroke Islands.

Amity Point: A sand bar across the mouth of the Brisbane River made it unnavigable to larger ships so, initially, goods were offloaded at Amity and then transferred to Brisbane. While visiting Moreton Bay in 1825, Edmond Lockyer records '…a large bark hut and two or three soldiers at Amity Point. Provisions from the ship "Lalla Rookh" had been landed there a couple of months earlier and were being taken gradually up to Brisbane in a smaller boat.'[63, 75]

The unloading and double-handling of goods at Amity was a tedious task. Governor Darling, visiting Moreton Bay in 1827 in HMS 'Rainbow', ordered that a depot be established at Dunwich (then known as Green Point), where there was a good anchorage and the plentiful freshwater supply of nearby Yerrol Creek.

The brig 'Amity', returning to Sydney from the Moreton Bay settlement in 1824, was the first ship to use the South Passage: 'John Gray had surveyed and buoyed the South Passage about April 1825, and from that time ships proceeding to the Brisbane River went past Amity Point and Dunwich around the southern side of Peel Island, across to the "Green Islands" [St Helena Island, Mud Island, and Green Island] to the river.'[100]

Following the transfer of the penal settlement from Redcliffe to the Brisbane River in 1825, the South Passage became the main shipping channel into Moreton Bay, and a pilot station was established at Amity. The pilot was provided with a seven-convict boat crew and three soldiers. The South Passage was used for shipping until the wreck of the 'Sovereign' in 1847, when 44 lives were lost. Following this disaster in what is still treacherous water, the pilot station was moved to Cape Moreton on Moreton Island.

Amity has always been faced with the 'problem' of foreshore erosion. Since the area was first surveyed in 1886 for residential allotments, more than 350 metres of land have been swept into the Rainbow Channel. The old pilot station is now well out in the channel, and Tom Welsby's house was one of many dwellings moved back to safety repeatedly from the encroaching sea. Welsby, a well-known historian, lived at and visited Amity for many years. His house was dismantled. The timber from it was used in the residential section of the kiosk 'Cabarita'.

HMS 'Rainbow', under the command of Captain Rous, was the first warship to enter Moreton Bay. The names of Rous and his ship are perpetuated on bay charts. Stradbroke was officially named 'The Isle of Stradbroke' in 1827 in honour of the first Earl of Stradbroke (Viscount Dunwich). The Rous, a deep channel, runs between South Passage and the Brisbane River mouth, and Rainbow Channel lies along the northwest shores of Stradbroke.[100]

Plate 57 Pilot Station, Amity Point. A watercolour by Captain Owen Stanley, 1847 (Mitchell Library)

Plate 58 'The harbour master's boats-crew in a mess, Moreton Bay'. Painting by Owen Stanley, 1847-1848. 'The pilot at Amity Point, with his crew of convicts, used to row through the breakers in the South Passage to meet incoming ships. This was hazardous work, and one pilot (John Tosh) and two of his crew were drowned in January, 1834.'[100] (Mitchell Library)

Plate 59 Captain Patrick Logan, commandant of the Brisbane settlement, established the first settlement at Dunwich in 1827. (Mitchell Library)

Plate 60 Tom Welsby maintained close association with the Aborigines of Stradbroke and recorded their history and many island anecdotes. (Oxley Library)

Plate 61 Tom Welsby's house at Amity, 1950s. The house was moved back from the encroaching sea many times. Eventually it was pulled down. Some of its timber was used to build the present kiosk, 'Cabarita', which is now a private residence. (Authors)

Dunwich: To establish the Dunwich depot, Captain Logan had military and convict barracks and a storehouse constructed near where the present vehicular ferry ramps are. Part of the foundations of the ferry approaches were built by convicts. Captain Logan's successor, Captain Clunie, writes in 1831: '...The great objections to Dunwich are the loss of stores it occasions, that it affords every facility for smuggling, that it does not expedite the unloading or reloading of vessels while it is attended with expenses, and employs some prisoners, whose labour is attended with little advantage to the public, and who, not being strictly looked after are in a certain measure defeating the object, for which they were sent to a penal settlement.'[75]

Captain Clunie also reported an Aboriginal attack in 1833 on the pilot station and blood spilt on both sides at the 'Battle of Kappembah'. The Aborigines' account of this 'battle' was recorded by Welsby in about 1880. It differs slightly from Clunie's. After the soldiers spent an uncomfortable day in the swamps shooting at phantom figures and dodging behind grass hummocks, and the Aborigines ran out of throwing weapons, a halt was called to the battle. Aborigines and soldiers all walked home to Dunwich together. This was a minor skirmish, one of the few altercations between the Aborigines and European settlers in their long contact.

Some agriculture was attempted at Dunwich. Cotton planted in 1828 was not a success. Allan Cunningham and Charles Fraser, the Colonial Botanist, visiting Dunwich, described the cotton plantation as neglected.

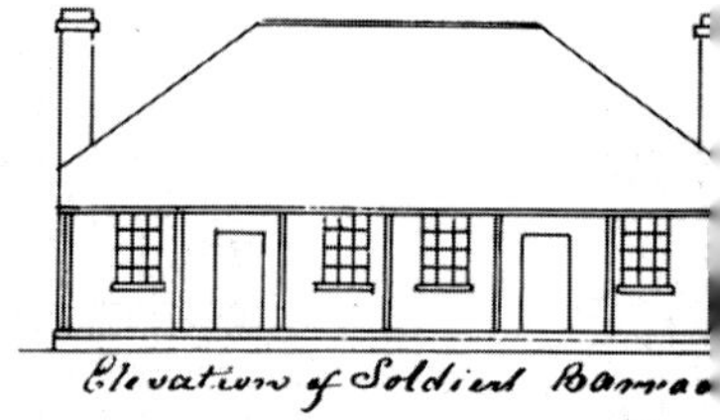

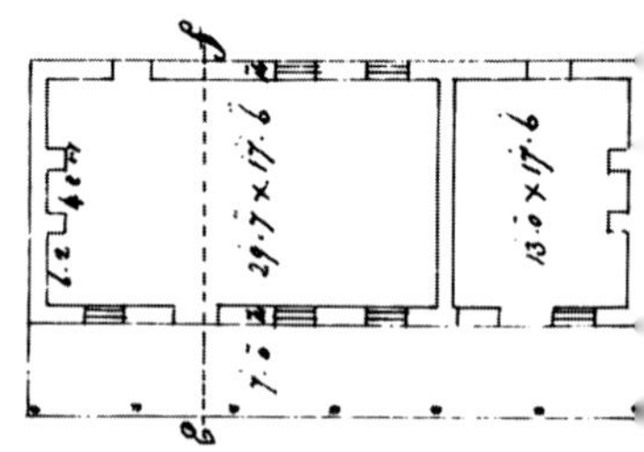

Fig 26 Plan and elevation of soldiers' barracks, Dunwich, late 1820s (State Archives: Moreton Bay plan 7)

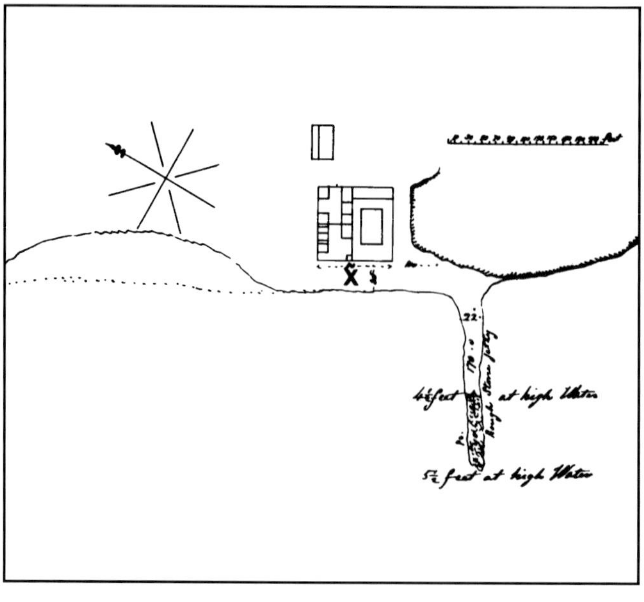

Fig 27 Early Dunwich: plan of jetty, store and barracks. The store was in the right-hand section of the compound. The building at the rear was the soldiers' barracks.[100] An X marks the site of the privy. Many island residents incorrectly believed what remains of the pit for this privy was a dungeon. The pit can still be seen on the Dunwich foreshores. (State Archives: Moreton Bay plan 7)

Fernandez Gonzales: 'A pure Manila man, with the strong passionate blood of the Spaniard, and truly with a strong passionate determined heart of one'[105] settled at Amity in the 19th century. Gonzales hunted dugong professionally for sale to other settlers. Welsby records Gonzales' estimate of natives living at Amity Point in the year 1853 as 'upwards of 150'. Gonzales married one of the Aboriginal women.

Escaped convicts: Nine hungry convicts escaping from Norfolk Island in 1853 landed on North Stradbroke, probably in the vicinity of Adder Rock. Circumstances at the time of the escape forced the convicts to take with them a Norfolk Island constable and coxswain. Leaving one convict to guard the constable and coxswain, the other eight convicts walked to Amity where they found Gonzales. Despite his hospitality, they stole his boat and stores of sugar, tea, flour and meat which he had recently brought from Brisbane. The Aborigines, who had hidden behind the tea-trees, came out in a body to stop the thieves. Unfortunately the convicts were armed. In the scuffle, one Aborigine was wounded and one convict taken captive. Next day, the convicts returned, rescued their captive mate, wounded several more natives, and stole two dugong that Gonzales had just retrieved from his nets on Pelican Bank. Two convicts were landed to retrieve the coxswain, constable, and their convict guard. Frightened by the natives' presence, the six convicts deserted their mates. They set sail for the bay, where they lived like kings on Gonzales' stores and dugong. They terrorised people in Moreton Bay for about 16 days until their recapture. They were sentenced to a further 15 years' imprisonment by the Brisbane Circuit Court.[27a]

Plate 62 Pencil portrait of Gonzales with his four daughters, 1891 (Oxley Library)

Plate 63 Fernandez Gonzales and his family, 1891 (Oxley Library)

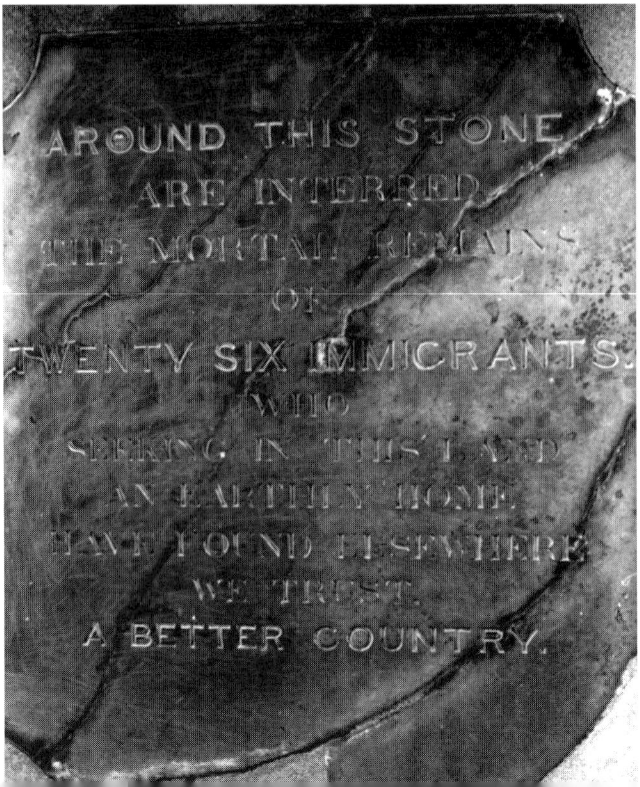

Plates 64 (left) and Plate 65 (opposite) Dunwich Cemetery, showing the graves of Doctors Ballow and Mitchell, along with those of 26 of their patients. All died in 1850, victims of a typhus epidemic. (Authors)

Quarantine: Moreton Bay ceased to be used as a penal settlement by 1839. By 1842, free settlers began to arrive. This meant the end of institution life in the colony generally. On Stradbroke, however, institution life continued. Captain Wickham of the Moreton Bay settlement was asked in 1849 to recommend a suitable site for a quarantine station. He chose Stradbroke Island because of its relative isolation and plentiful water supply. To ensure complete isolation, it was suggested that the entire island be used. However, in July 1850, only the northern portion of Stradbroke was proclaimed for quarantine purposes.

In September, 1850, the 'Emigrant' carrying 64 typhus patients arrived at the newly declared and unprepared Dunwich Quarantine Station. There was no suitable housing for the patients. The old government barracks had fallen into disrepair and there were only 12 small worn, leaking tents. Makeshift shelters were rigged from pieces of sail and ship's canvas, but these were little protection from frequent storms. Many patients suffered recurring fever and died. Along with 26 of the 'Emigrant's' passengers, the ship's surgeon, Dr George Mitchell, died of typhus. A headstone in the Dunwich Cemetery tells the tragic story: 'Around this stone are interred the mortal remains of 26 immigrants who, seeking in this land an earthly home, have found elsewhere, we trust a better country.'

Dr P W Mallon, of Brisbane, volunteered his services but soon contracted the fever. A second Brisbane volunteer, Dr Ballow, Brisbane's resident surgeon, died of typhus and was buried beside Mitchell in the Dunwich Cemetery. Ballow Chambers on Wickham Terrace, Brisbane, Ballow Park, at Dunwich and Ballow Street, Amity are named to recognise his contributions. Mallon recovered to tend the last of the typhus patients.

Benevolent Asylum: '...The Benevolent Asylum had been established at Dunwich in 1864 in the quarantine buildings, and moved out temporarily shortly afterwards and again in 1865 because the place was required for quarantine...old and infirm paupers and some young persons who were badly disabled, were sent there, and later inebriates as well, and for a time from 1892 there was also a lazaret.'[75]

The lazaret and quarantine stations were later relocated on Peel Island. The move was made after the postmaster's son, Laddie Agnew, contracted leprosy. In the beginning, the government did not extend itself beyond providing the bare necessities of life.

Benevolent Asylum
Dunwich
6th September, 1915.

The Under Secretary,
Home Department,
Brisbane.

Sir,
I have the honour to inform you that the increased cost of meat viz. 6d. for beef and 7d. for mutton as notified to me this month, will increase our expenditure in this direction to an enormous degree...£2,000 a year.

Under the above circumstances, I consider that it would be advisable to slightly decrease the issue of meat at present allowed to the inmates, and to make up for such loss with a less expensive form of diet...

I have the honour to be
Sir,
Your obedient servant

L.E. Row
Medical Superintendent

Over the years, conditions at the government-run Benevolent Asylum improved. The postmaster and his son Laddie gave gramophone concerts. Visiting concert parties entertained the inmates and, later, movies were screened three times a week in the Victoria Hall. The institution boasted a well-stocked library. Dunwich was serviced from Brisbane by the Government steamer 'Otter' of 222 tonnage. Later, the launch 'Karboora' ran an additional service from Cleveland.

Opposite: Plate 66
A reporter from 'The Queenslander' recorded his impressions of the Dunwich Institution in 1895: 'There are, it is true, some painful things to see. The maimed, halt and blind are piteous objects everywhere and a community almost wholly composed of the infirm and aged, with little or no young life about to gladden it, suggests a world all winter without promise of spring. During 1894 the Dunwich people manufactured a considerable quantity of clothing as well as beds and bedding, those of the inmates who are not quite incapacitated being kept at work as much as possible, both to keep them occupied and to help make the burden of their support fall lighter upon the general community. This custom has another beneficial result...it keeps out to some extent those able-bodied loafers who are always seeking to obtain admission and live a life of ease at the expenseof their fellow colonists.The net cost of maintenance for the year was £13 per inmate.'
A staff artist presented a more pleasant, tranquil picture of the place.
(Oxley Library)

Plate 67 Visitors to the Dunwich Institution (Oxley Library)

The institution was costly and difficult to administer for many reasons. Repeated recommendations for its removal to the mainland were disregarded until matters came to a head in 1947: '…The Dunwich Institution was transferred in haste, caused by a strike aboard the government steamer "Otter", when the crew refused to sail the ship because the water was leaking on their never-used bunks. Premier Ned Hanlon was so incensed by this work stoppage that he set about buying the old R.A.A.F. Sandgate Station that was on the market for the ridiculous price of twenty five thousand pounds. The quoted price to replace the steamer "Otter" was in the vicinity of two hundred thousand pounds.'[32a]

The 768 inmates were transferred to Eventide at Sandgate in 1947. The Sandgate site was stark and bare compared to Dunwich with its green grassy slopes and magnificent old trees planted by former doctors, institution officers, and their wives.

The removal to Sandgate of the Dunwich Benevolent Asylum closed a chapter in the history of Stradbroke. It was the end of an institutionalised island. Stradbroke was now sought out by those seeking work and recreation in a tranquil island setting.

Fig 28 St Mark's Church of England, Dunwich, was constructed in the days of the Benevolent Asylum.

Opposite: Wreck of the 'Rufus King'

Plate 68 Mrs P Agnew, wife of the Dunwich postmaster, with visiting Church of England Bishop at Dunwich in the 1920s. Mr Agnew and his son Laddie brightened the lives of the inmates of the institution by organising concerts. (Oxley Library)

Shipwrecks

Five or six well-known large vessels and many smaller fishing craft have come to grief on the sand bars which surround North Stradbroke Island. South Passage Bar has figured prominently in these disasters, most of which have occurred during squalls when visibility has been bad. On two occasions, the captains confused Point Lookout with Cape Moreton. No large vessels have run aground near Stradbroke since 1942.

Spanish galleon: Several islanders (including an employee of one of the sandmining companies and a stockman working on the island) claim they have seen in the Eighteen Mile Swamp the remains of a vessel believed by some to have been a Spanish galleon. Ted Ruska, who worked for Billy North, told his children he saw the ship with its masts out above the marsh when he was a boy. Local historian, Bonty Dickson, has written of this ship: 'The year of the shipwreck is not recorded, but several people living today swear of its existence or part of the vessel embedded in the swamp or sand two miles north of Jumpinpin. A lot of people confuse the exact area of Jumpinpin with Swan Bay and, as the place the bar broke through was originally two miles south of where it is today, it could cause confusion as to the location of the wreck.'[32]

He also records local reports of the existence of an old Spanish coin held by an island resident.

The possibility that a Spanish ship, hopelessly blown off course on its way to South America, and wrecked on the sand spits at the south of what is presently North Stradbroke, cannot be discounted. Historical studies suggest that, although Lieutenant James Cook was the first to report eastern Australia to the British government, he did not 'discover' eastern Australia. Spanish and Portuguese ships travelled through the Torres Straits and may have moved along the east coast at least 200 years before Cook; and Chinese fleets are known to have charted sections of the Australian coastline at least as early as the 15th century.

Several facts, however, militate against the existence of a wrecked Spanish ship on Stradbroke.

> There are no relics of the ship to be seen or examined by experts. (It is, of course, possible that these have been destroyed by bushfire.)

> Opinions of those who claim to have seen ship remains near the south of the island differ markedly regarding location – eastern vs western side of the swamp. Also, sight records date from the 1940s and have not been confirmed since.

> The Spanish coin alleged to have come from the

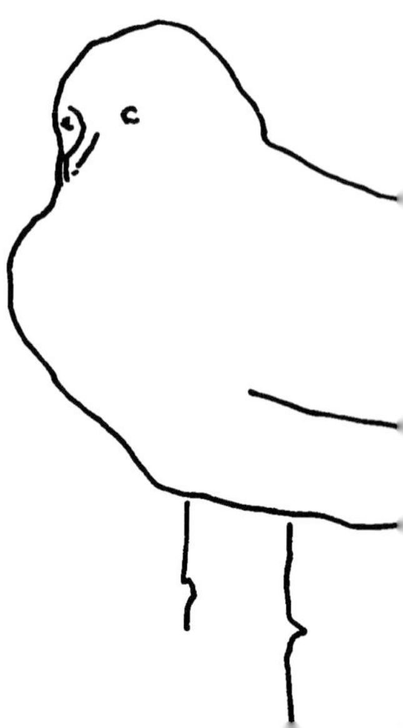

wreck cannot be located, and further, even if such a coin exists, it may not have originated on a Spanish galleon. Spanish coins were issued by British banks in 1797 for use on British ships and were common currency on the Pacific trade routes.

'Sovereign': The 'Sovereign', a paddle steamer of 119 tons, was wrecked in March 1847 on the South Passage Bar. Forty-four lives were lost. The Aborigines rescued ten survivors, one of whom was the captain.

After sheltering at Amity until southerly gales abated, Captain Cope, under full steam, headed out through the South Passage separating Moreton from Stradbroke. Crossing the bar, the ship rode the first two breakers, but on the third wave the engine failed. The captain attempted to anchor. The battering waves broke the ship's paddle boxes.

'The rollers now broke upon the devoted vessel with great violence, carrying away bulwarks and causing the wool and billets of wood to move violently about the decks, whereby three men were killed, while several more had their arms and legs broken, or [were] otherwise disabled. The captain then told the passengers that he saw no hope of saving the vessel, as she was still dragging towards the spit...

'No pen can describe the awful scene that presented itself on board at this time. The passengers were in the utmost consternation, they set up most piteous cries for help; some ran to the side, and in the agonies of despair, plunged into the sea, in the hope that they might reach the shore in some way or the other.'[105]

The southern beach of Moreton was littered with wool and splintered pieces of wood. The bodies of Mrs Gore and one of her children were washed up on the beach. As the tide was ebbing, most of the other bodies were carried out to sea. It was revealed after the tragedy 'that the ship's timbers were originally defective, the fastings greatly weakened by repeated repairs and the number of tree nails and bolts used, accounted for the hull breaking up before she grounded... [T]he engines were salvaged from a wreck before being installed in the 'Sovereign'... [T]hey had been under water for some time before they were salvaged.'[81]

'Countess of Derby': After the disastrous wreck of the 'Sovereign', ships' masters preferred to enter Moreton Bay by way of Cape Moreton. The captain of the 'Countess of Derby', choosing to disregard the fate of the 'Sovereign', wrecked his ship on the 31st October, 1853, about a mile inside the spot where the unfortunate 'Sovereign' was wrecked in 1847.

An account of the disaster appeared in the 'Moreton Bay Courier' of Saturday 5th November 1853: 'We regret to have to report the loss of the Countess of Derby, bound from Sydney to Cleveland, whilst bearing up for the latter place through the South entrance of Moreton Bay on Monday last. The Countess of Derby, barque, 329 tons, Starcich, commander, left Sydney on the 14th ultimo, having been chartered by Messrs. Monteifiore, Graham, and Co., to convey wool and other produce from Cleveland to London. The vessel appears to have made a long passage to Point Lookout, off which place she arrived on Monday last, the wind being N.E. The southern entrance being the nearest to Cleveland and the Captain being desirous of saving the daylight, he brought the vessel in there instead of the usual entrance at Cape Moreton and she soon struck on the South spit.'

Fortunately no lives were lost in the wreck. The following notice appeared in the 'Moreton Bay Courier', Saturday November 12th 1853:

SALE BY AUCTION
On account of whom it may concern.
The wreck of the Countess of Derby, as she lies stranded on Stradbroke Island, with masts, rigging sails, and anchors, chains etc., etc., etc.
For Sale by Public Auction
on Tuesday, the 15th instant,
at 12 o'clock,
At Mr. Watson's Wharf, Kangaroo Point
Also Ships' Long-boat, Jolly Boat, and gig, with oars, sails, etc.
TERMS CASH

This was followed, a week later, in the same paper, by: 'The wreck of the Countess of Derby, as it lies at the South Passage, was sold by auction last Tuesday for £29. The boats and other articles saved from the wrecks, realised £140 more.'

'Phoebe Dunbar': In 1856, the captain of the 'Phoebe Dunbar' mistook Point Lookout for Cape Moreton. His ship, carrying 270 immigrants, was grounded at Amity Point; only one crewman was lost. The ship was repaired and re-floated. The wreck of the 'Phoebe Dunbar' hastened the completion, in 1858, of the Cape Moreton lighthouse.

'Cambus Wallace': The barque 'Cambus Wallace', a steel ship of 1651 tons, was wrecked in September 1894 on the narrow part of Stradbroke where only a line of dunes separated the ocean beach from Swan Bay. Within two years, following a storm, the sea broke through into Swan Bay, resulting in the separation of North Stradbroke from what has become South Stradbroke Island. The Glasgow-built barque was on her maiden voyage. Throughout the trip to Brisbane, she had been constantly subjected to gales and bad weather. When off Stradbroke, the captain believed he was some 36 miles off the land.[105] Squalls reduced visibility. The ship struck the line of surf. In a few hours, the sea broke up the vessel. Of the crew of 22, the steward, the cook, three able seamen and the carpenter were drowned. The remaining men managed to struggle to the beach. Five of the dead were buried on the sandhill overlooking the site of the wreck. Wood crosses and a rail were later erected to mark the graves. When the sea broke through into Swan Bay in 1896, the graves disappeared. The sixth man, Peoples, in poor shape after reaching shore, died on his way to Brisbane and was buried in the Toowong cemetery.

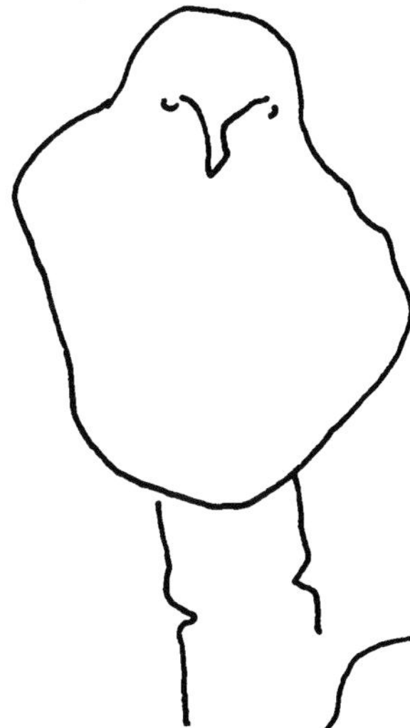

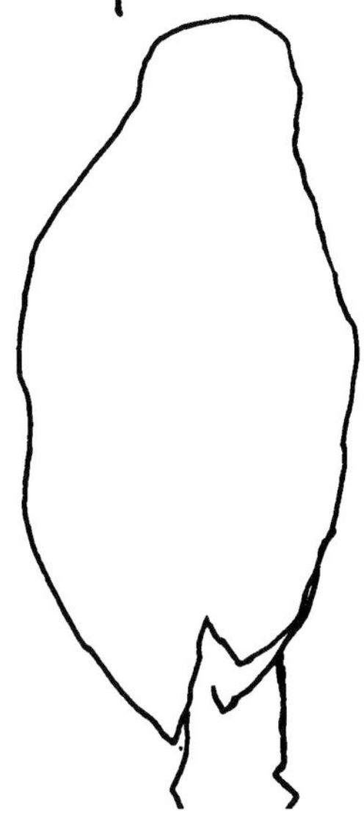

Among other things, the ship carried a cargo of spirits, beer and dynamite. To protect the goods during salvage operations, customs officials were called in, and police patrolled the beach. Excessive activity in the area during salvage operations destroyed fragile stabilising dune vegetation. This disturbance coupled with the authorities exploding the dynamite, may have hastened the breakthrough by the sea in 1896.

Ben Cropp, an underwater expert, diving on the submerged wreck many years later salvaged some of the ship's cargo of whisky which he described as 'barely drinkable but highly potent'.

'Prosperity': At night during rough weather, the 133 ton brigantine 'Prosperity' from Sydney was wrecked in 1902, 800 metres southeast of Point Lookout. The captain (P J Peterson) and four crewmen managed to reach the safety of the shore. The cook and mate were drowned. On one of the dunes of Deadman's Beach, the skeleton of an adult male and the remains of a leather lace-up ankle boot were partly uncovered in 1956. This suggests that one of the unfortunate crew was buried there. The 'Prosperity' carried a cargo of sugar machinery for Mourilyan Harbour in north Queensland. Pieces of rusted machinery can still be seen wedged in the crevices of South Rock.

Mr North, who ran cattle on North Stradbroke to supply beef to the Dunwich Institution, had a stockman's hut at Point Lookout at this time. North advised the authorities of the wreck. He also kept the beach free of strangers and cared for the five survivors in his hut where he had an ample supply of food.

Plate 69 Survivors of the wrecked 'Prosperity' (Rahnsleben scrapbook, Fryer Library)

Plate 70 Bill North's hut, Point Lookout. Here, Bill North cared for survivors of the 'Prosperity'. (Wendy Goebel)

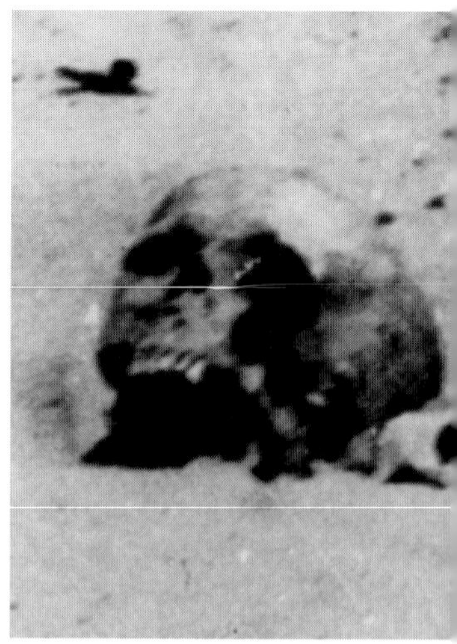

Plate 71 A grisly sight uncovered on Deadman's Beach in 1956. The leather boot and skeletal remains of an adult male are reputed to be from one of the crew of the 'Prosperity'. (Rahnsleben scrapbook, Fryer Library)

Plate 72 All that was left of the 'Prosperity' one year after it was wrecked in 1902, near Point Lookout. Photographed in South Gorge, in 1903. (Oxley Library)

Plate 73 Ted Ruska with the dictionary from the 'Prosperity', many years after the wreck. Ted Ruska and Bill North assisted survivors. (Authors)

Plate 74 Mast of the 'Prosperity' on Main Beach, January 1956 (Joan Gregory)

Above: Plate 75 'Rufus King' breaking up on the South Passage Bar, in 1942 ('Courier-Mail')

Below: Plate 76 The hospital ship 'Centaur' was torpedoed by Japanese in 1943 near Brisbane, with the loss of many lives. (Oxley Library)

'Rufus King': In July 1942, the Brisbane 'Courier-Mail' described the war-time wreck of the 'Rufus King' on the South Passage Bar (where the rusting stern of the ship lies below the water line).

'In drizzling rain, the American liberty ship "Rufus King", bound from Los Angeles to Brisbane, had gone hard aground – just 72 days after her keel was laid in an American shipyard…On board were enough medical supplies to outfit three complete field hospitals, nine Mitchell bombers in crates, and stores of aviation fuel…Twelve hours after she went aground the sea broke her in two as rescuers tried to take off the crew. The bows began drifting…The crew of 30 or 40 were landed without a casualty…The bow was sealed and towed to New Guinea, where it finished the war as a floating workshop…The captain of the "Rufus King" was taken back to America under arrest.'

At the subsequent inquiry, the captain claimed that he mistook Point Lookout for Cape Moreton.

'Centaur': The hospital ship 'Centaur' was torpedoed in 1943 by a Japanese submarine, 40 miles off Brisbane. The ship was on her way to New Guinea to bring back wounded. She carried nursing sisters, medical orderlies and doctors. Of the 332 persons aboard, only 64 survived the disaster, which 'Truth' of 17th October 1954 describes:

'Her white hull glistened under arc lights, which made the huge red crosses and the green (non-combat) band visible for miles…Just before dawn, "Centaur" sailed into the sights of a lurking Japanese submarine. A torpedo struck her amidships, buckling her plates and breaking her in two. In a horror of flames, crashing rigging, whirlpools, and sharks, 268 men and women perished…"Centaur" sank in three minutes, before her radio could signal the atrocity ashore…Pounded by tropical rain, the 64 survivors clung to wreckage and emergency rafts for 36 hours. They had almost abandoned hope, when a plane dipped through the clouds and brought a destroyer to the rescue.'

The site of the final resting place of the 'Centaur' has never been confirmed. The fact that in the early hours of October 1943, the Durbidge family resident at Point Lookout heard a loud explosion, and Harold Wale saw out to sea a glow light up the sky, suggest that the 'Centaur' lies off Moreton Island.

Point Lookout lighthouse: The automatic Point Lookout light flashes three times in every 15 seconds. The light is brought into operation whenever necessary by a sun valve. The light is 72.5 metres above sea level and visible from 18.6 kilometres at sea.

The lighthouse, built in 1932, was operated on acetylene gas. Construction materials and acetylene cylinders were off-loaded from the 'Cape York' to a couple of work boats which were towed by a motor launch to just outside the surf break. The work boats ferried the materials inshore to a small temporary jetty constructed on Cylinder Beach. A small hut to store the cylinders was built in the eastern corner of the bay. Its cement base is still there.

Once during the ferrying operations, the launch ventured too close to the surf where it capsized and sank. A hired light aircraft failed to locate the sunken launch. Unloading operations were shifted to the calmer waters of Amity Point. Ernie Odlem transported the materials to Point Lookout by truck along the beach at low tide.

Ron Stubbins, one of the men employed on the light construction, says that water for cementing the light base was pumped in two stages from the spring at the head of South Gorge. Mr Stubbins supplied the photographs associated with the light's construction. A camp for the construction gang of about 20 men was set up on the north side of the valley, near the site of the present Samarinda units development.

The tower and dome were originally surrounded by a galvanised iron fence which has been removed. The lighthouse was painted white with a red dome. At the official inquiry that ensued after the 'Rufus King' ran aground on the South Passage Bar in 1942, the captain claimed he mistook the red and white Point Lookout light for the Cape Moreton light which was also red and white. To avoid further confusion, the light at Point Lookout was painted white only, and the Moreton lighthouse painted with alternate red and white bands.

When in 1988 electricity replaced acetylene gas to operate the light, the 1932 building was modified. The surrounding galvanised wall, the external steel ladder, the dome, and the light prism were removed.

Fig 29. When in 1988 main electricity replaced acetylene gas to operate the light, the 1932 building was modified. The surrounding galvanised iron wall,external steel ladder, dome, and light prism were removed. The prism is housed at North Stradbroke Island Historical Museum at Dunwich.

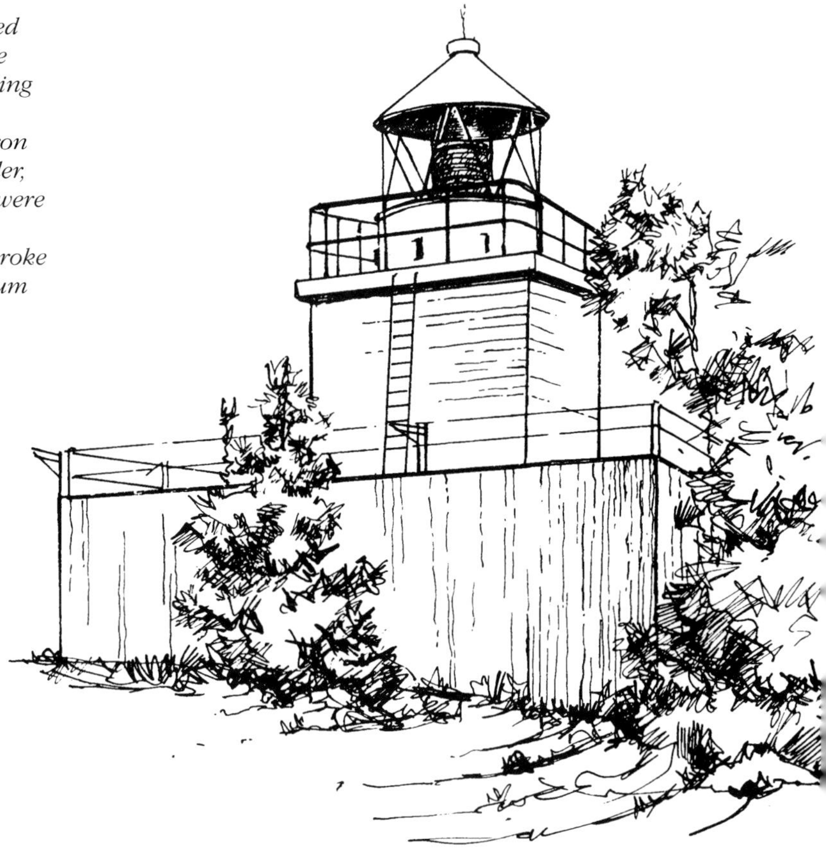

Plate 77 In 1825, Major Lockyer recommended that a lighthouse be constructed at Point Lookout. Many years and shipwrecks later, the Point Lookout lighthouse was built in 1932. Building materials and gas cylinders were ferried ashore at Cylinder Beach. (Rahnsleben scrapbook, Fryer Library)

Plate 78 Building materials and acetylene gas cylinders being loaded into a pontoon, Cylinder Beach, 1932 (Rahnsleben scrapbook, Fryer Library)

Plate 79 Men employed to build the lighthouse were housed in a tent camp on a site near the present Samarinda units. The freshwater spring that bubbled up in the valley below the camp and the Banksias that sheltered the camp are now buried under a sealed car park. (Rahnsleben scrapbook, Fryer Library)

Industry and Development

Plate 80 Hauling in a full net, Main Beach, early 1950s (The late Bob Pattison)

Left: Fig 30 Tailor (Pomatomus saltator) are important to both commercial and recreational fishermen. The peak period is during winter when the fish migrate into southern Queensland waters from New South Wales. They are netted commercially, and line-caught by anglers from ocean beaches and in estuaries. The value of the commercial fishery in 1979-1980 was about $209,000.[118] In 2002-2003, the estimated value was $100,000. Because of the importance of tailor fish to recreational fishermen, a quota is imposed on commercial catches.

Right: Fig 31 Mullet (Mugil cephalus) is the most valuable finfish in southern Queensland. Its estimated value in 1979-1980 was $920,000.[118] The 2002-2003 estimated value was $900,000.

Fishing: 'A century ago, 18-footer boats raced from North Stradbroke to be first at the Brisbane market with the mullet catch. It launched a new working man's sport – 18 foot sailing and racing.'[24a]

Since The Dreamtime, mullet have left coastal rivers and moved to the open sea to spawn. Vance Palmer describes the gigantic, frantic schools of fish on their winter migration, '...sluggish with the weight of roe, a mindless mass of spawning life'.[78]

The professional net fishermen wait on the beaches to run their nets round the seething masses. Migrating tailor follow a little later in the season. Loaded nets are no longer manhandled but are hauled up the beach by tractors. The fish are loaded into trucks waiting on the beach. Bay net fishermen operate on different lines. Equipment usually consists of a motor boat towing several net dinghies.

Bob Durbidge and his family were the first to work from the Point Lookout ocean beaches to fish the adjoining reefs professionally, in the 1940s. During the war, the army requisitioned his Townsville-based mackerel boat, 'Aroetta', so he shifted his business and set up camp at Cylinder Beach. From there he worked with one of the 'Aroetta's' heavy wooden work boats powered by a Blacksland Pup 5.6 hp.

Where once ten professional fishermen worked their boats from these beaches, only a couple still fish for Spanish Mackerel and reef fish. It is mainly recreational fishermen and women that now fish from the beaches. The wooden boats have been replaced with light streamlined aluminium craft equipped with fast powerful motors, walkie-talkie radios, and echo sounders.

Plate 81 Durbidge's ice box fish storage, Adder Rock, 1940s. (Joan Gregory). As there were no refrigeration facilities and no power at Point Lookout, the Brisbane Fish Board delivered slabs of ice packed in sawdust-filled hessian bags to Hayles Wharf at North Quay. Hayles Bros off-loaded ice at Amity. From there it was taken to the Point Lookout ice-box where fish were stored ready for transport to Brisbane.

Plate 82 Medal won by Bill North for fish from his cannery at The Two Mile. In addition to canning surplus fish on a seasonal basis, Bill North ran cattle which he sold to the Dunwich Institution. (Wendy Goebel)

Fig 32 Whiting (Sillago sp) are caught from the beaches all year, especially in northerly weather.

Oystering: European settlers quickly realised the commercial potential of the extensive inshore oyster banks which had been used as food by the Aborigines for centuries. The Moreton Bay Oyster Company was formed in 1876. As well as cashing in on the Aborgines' traditional oystering grounds, owners of the company were astute enough to take advantage of local knowledge and expertise. The company employed many Aborigines and people of Aboriginal descent.

The Moreton Bay Oyster Company extended its operations from Southport to Maryborough. Boatloads of oysters were transported to the markets. A 70-foot oyster boat, 'Sir Arthur', built at Newstead, was named after Sir Arthur Palmer, chairman of the Moreton Bay Oyster Company. Norman Wright, Snr, the well-known boat builder, is quoted as saying he 'would hazard a guess that she could carry about 400 bags of oysters which were taken from banks south of Wallum Creek to Couragee. More oysters were taken from this channel than from any other oyster banks in Queensland waters and...were exported to as far as Perth.'

The development of the northern New South Wales industry, and a disease that depleted the Queensland banks, resulted in the ultimate winding up of the company in 1956. Bonty Dickson, a latecomer to the industry, bought the old dredge 'Hercules' for experimental oyster culture. Unfortunately the venture was not as successful as hoped. Insufficient sunlight inhibited the oysters' growth.

Oystering on Stradbroke is now carried on only by individual operators rather than on a company basis. Many island families have worked island and adjacent banks for generations.

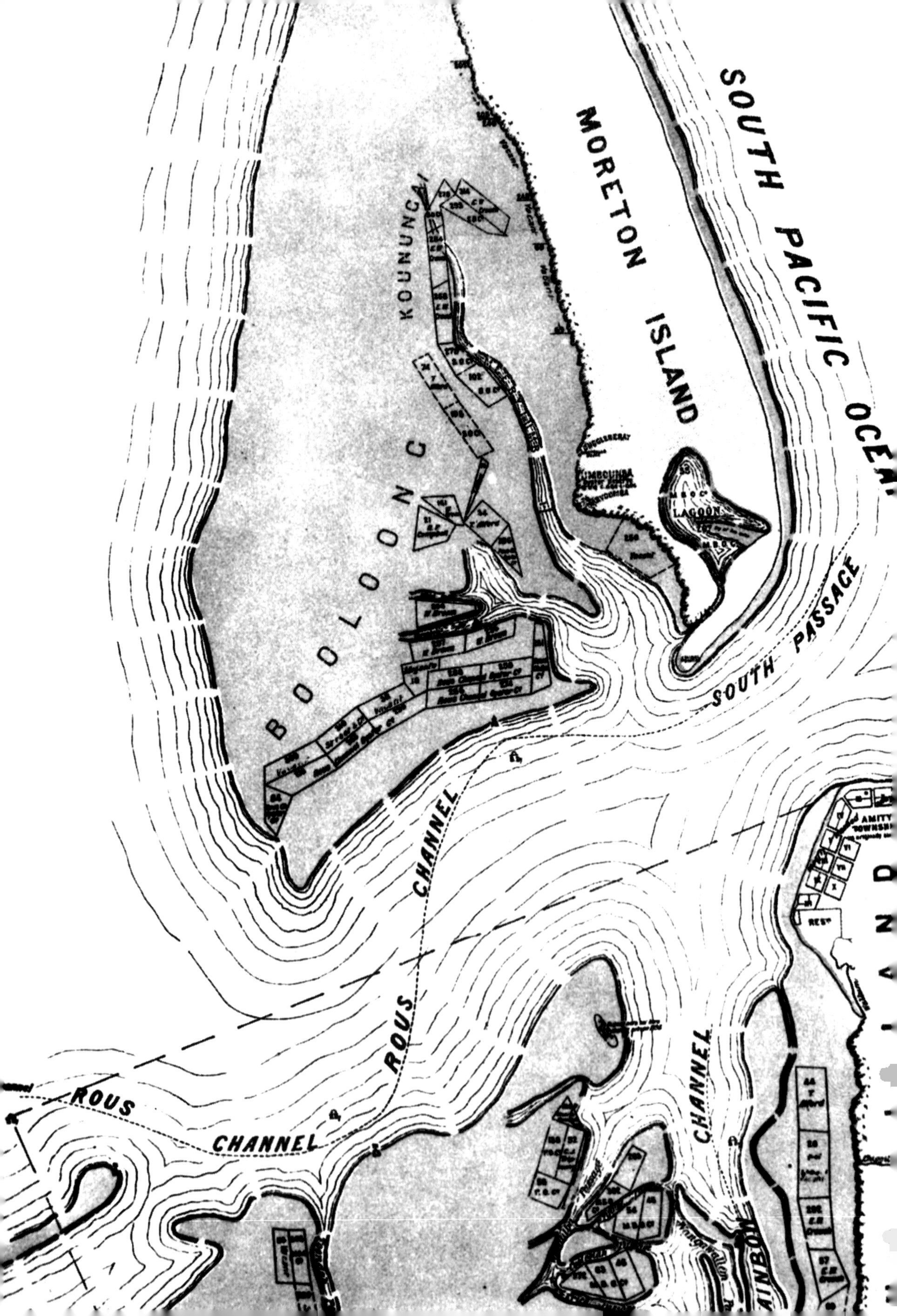

SOUTH PACIFIC OCEA
MORETON ISLAND
KOUNUNGAI
BOOLOONG
LAGOON
SOUTH PASSAGE
ROUS CHANNEL
ROUS CHANNEL
CHANNEL
AMITY

Forestry: Stradbroke has only very limited timber reserves when compared with other southeastern Queensland sand masses on which rainforests are extensive (eg, Fraser Island and Cooloola).

Some Cypress Pine and hardwood have been cut on Stradbroke since the 1940s when Dave Thurecht cut island cypress. The logs were loaded onto punts at the Amity Swing Basin and near the Amity Bar, from where they were transported to Redcliffe for milling. In the 1950s when an island timber lease was granted, a mill was constructed at the Two Mile, Dunwich. Because the mill was poorly set up, it was uneconomic to operate. (Greg Litherland, personal communication)

Associated Minerals Consolidated was granted a lease of approximately 965 hectares east of Dunwich to plant Slash Pines (*Pinus elliotti*). Although the mineral sands company planted the pines with a view to giving something to the local population in return for mining the island, the project was not environmentally friendly. Introduced pine forests support fewer animals than adjoining native habitats. Those animals that can use them for food and shelter are, almost invariably, common, wide-ranging species. Highly specialised animals cannot survive in them because they do not provide suitable habitats owing to structural simplicity and impoverished plant species diversity. Also, the pines spread to the adjoining native bushland. In 1995, fire partly destroyed the plantation; the burnt logs were sold by the State.

Greg Litherland has locally marketed island cypress pine and hardwood milled at Amity since 1971. Where once a timber lease permitted logging on State land, now logging on Stradbroke is restricted to a salvage operation where only logs felled prior to the mining of leases are available for milling. The sand-mining company has recently moved into its last big lease, Enterprise. With the Amity sawmiller restricted to taking timber from mining paths (the majority of trees in the path of mining operations are burnt), commercial milling on Stradbroke will cease. Also, the State has chosen to ignore the need to replant the burnt-out pine forest with native species to inhibit weed intrusion, and to cull the slash pines that have escaped into native bushland.

Sandmining: Stradbroke's mineral sands originated in the weathered volcanic rocks of the eastern high-lands of northern New South Wales. Minerals such as ilmenite, rutile and zircon are transported to the sea by the rivers draining the plateau, carried north by long

Opposite page: Plate 83
Oyster leases in 1884 when the Moreton Bay Oyster Company flourished (Oxley Library)

shore currents, and gradually deposited with beach sand. The accumulations of these minerals in the sand deposits of northeastern New South Wales and southern Queensland are amongst the world's richest.

'The deposits of black sand on the beaches of eastern Australia first attracted commercial attention about 100 years ago...The extraction of industrial zircon at Byron Bay in 1933 marked the beginning of modern beach sand mining on the east coast.'[18]

The demand for ilmenite and rutile which produce titanium (for use in top-quality white paint and high-speed aircraft) and zircon (for flashbulbs, uranium reactors, ceramics, and dyeing) during the war established the industry. Since that time, with increased demand for use in high technologies, the value of these minerals has continued to increase. Silica sand, used in making glass, is also shipped from the island.

The first shipment of mineral sand from North Stradbroke was made in 1950. What started as a labour-

Plate 84 Sandmining commenced as a small, labour-intensive industry on Stradbroke in 1947. Highly mineralised sand was shovelled by hand into 4 wheel drive army blitzes. (Authors)

Plate 85 Belt loading facilities, Dunwich. Mineral is now transported from the island on special sand barges. (Authors)

intensive industry when black seams of wave action concentrate with an 80% mineral content was shovelled by hand into old army blitzes and 4 wheel drive trucks working the ocean beaches, is today a capital-intensive industry. Highly sophisticated machinery is currently profitably handling 0.7% mineral content. Mining has shifted from the heavily mineralised beach-front dunes to the high transgressive dunes where minerals occur in reduced concentrations, but are still retrievable economically. The wet mining method is presently used on the island. Floating in a man-made pond, a dredge sucks the sand to a concentrating plant. The tailings or waste are pumped out behind the dredge.

In 1973, Consolidated Rutile produced minerals worth $4,115,058. For the year ending June 30th 1980, the same company produced $19,421,100 worth, following a production boost and higher prices. In 2002, $88.1 million worth of mineral was produced, and in 2003, $61.8 million.

Plate 86 By the mid-1970s, sandmining was a vastly different operation. Sand containing 2-3% mineral could be handled economically due to technological advances. In 2003, a 0.7% content is still economically viable to mine. (Bruce Cowell)

Above: Plate 87 Bucket loaded with mineral sand on the Herring Lagoon-Dunwich ropeway (The late Bob Pattison). The mining benefits have been won at the expense of the environment.

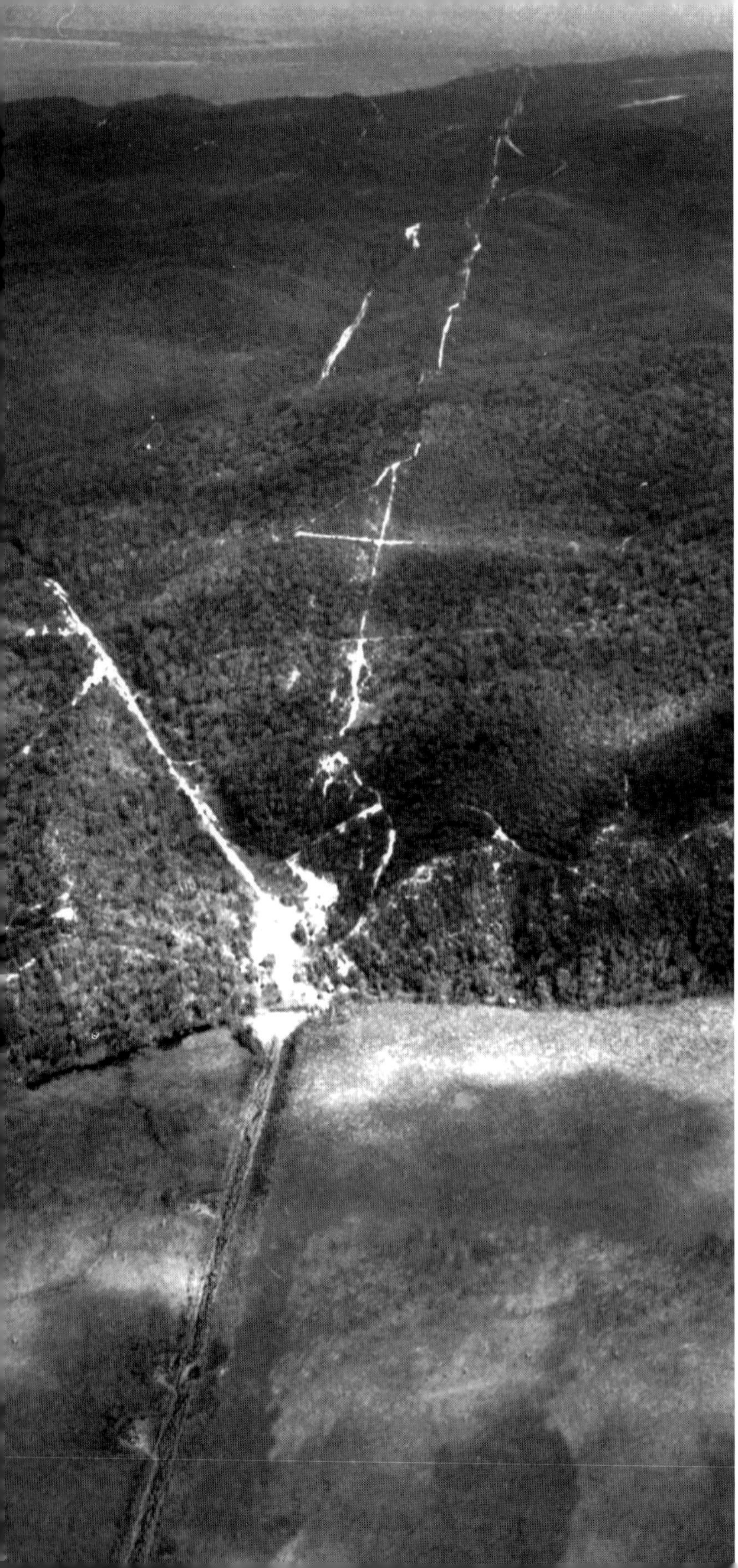

Left: Plate 88 Beginning of the 11 km aerial ropeway used to transport mineralised sand from Herring Lagoon to Dunwich, late 1950s. At that time, the ropeway was one of the longest in Australia. (Bruce Cowell).

Plate 89 The 'Myora' unloading mineral sands truck, Dunwich, 1981 (Authors). Mineral is now carried in bulk in sand barges to Brisbane for final separation.

Transport: When Bert Clayton commenced a regular beach vehicular service in 1934 from Amity to the guesthouse he was building on the present site of Samarinda units at Point Lookout, he opened up the east coast of Stradbroke.

In 1935, a north Queensland firm, Hayles Bros, built a kiosk at Amity, and commenced a launch service from North Quay, Brisbane. MV 'Mirimar', a 96-foot vessel, was the first launch on the Stradbroke run. MV 'Mirabel', of 60 feet, commenced additional services in 1937, and the 90-foot MV 'Mirana' which operated from Manly joined the fleet in 1947. Hayles were frequently committed to the expense of replacing jetties on Amity's eroding foreshores. At times there was no jetty. Supplies and passengers were ferried ashore in a small dinghy, or at low tide, a plank to the beach was used.

Varying beach conditions soon forced Clayton to provide a more reliable route for his transport service. In the late 1930s he built a road which was no more than two wheel tracks along the narrow high ridge overlooking the Amity Swamp. A trip in the old truck that served as a bus was scenic, always adventurous, and at times hair-raising. The narrow track was barely wide enough to accommodate one vehicle. When Clayton and an opposition operator from Amity had a disagreement, Clayton blocked the narrow track to his rival. The road, later reconstructed on its present alignment along the edge of the swamp, was known as the 'Sand Track' until it was renamed Clayton's Road to honour the man who built the first Amity-Point Lookout road.

Above: Plate 90 The only vehicle access to Point Lookout the 1930s was along the beach from Amity. Bert Clayton's bus was a T-model Ford. The woman sitting above the back wheel of the truck is Gladys Rahnsleben. (Gladys Rahnsleben)

Below: Plate 91 MV 'Mirabel' in the late 1940s at Amity. Erosion of Amity's shores had carried the jetty away, so goods and passengers were landed straight onto the beach. The woman closest to the camera is Amelia Campbell, and she is speaking to Doug Campbell. (Authors)

Dunwich and Amity were connected by road in the 1930s. A road was constructed under the government-sponsored Public Estate Improvement Scheme, via Blue Lake. For many years this and the Clayton track were the only road links between the three towns. The shortest and quickest way to Dunwich from Amity was to walk along the tidal flats at low tide. From Dunwich, a well-worn Aboriginal walking track terminated near North's cattle dip at Adder Rock, on the east coast. This track is no longer defined because of subdivision and lack of use.

During the war, parts of Stradbroke were declared restricted for defence purposes. An Australian Army permit was required to enter these restricted areas. The 6th Platoon US Army, and later the RAAF, operated a radar station at Point Lookout. Clayton's Point Lookout guesthouse (now the site of the new Samarinda units) was taken over by the services, and Bert Clayton then discontinued his bus service because there were, of course, no guests. Immediately after the war Clayton started a similar tourist venture at Cylinder Beach, at the present site of Cylinder Cove development.

No transport service operated between Amity and Point Lookout during the war, when Bob Durbidge began fishing on the reefs near Stradbroke. To overcome his own problems of transporting stores and fish, he bought an old truck. This became the first vehicle

Plate 92 Members of the 6th Platoon, US Army at Point Lookout during the war (Authors)

Plate 93 Durbidge's red bus on the ascent on the main road that ran round the headland to the Hotel Stradbroke, late 1960s (Authors)

Above: Plate 94 Terry Daley's view of the Friday night trip of the Point Lookout surf lifesavers from Amity to Point Lookout, 1950s (Authors)

Below: Plate 95 On the way to Point Lookout, 1950s. Tony Durbidge's ex-army 4 wheel drive Marmon Herrington on the Amity-Adder Rock beach. (Authors)

of the Red Bus Service. After the war, as the roads were still no more than two soft sand wheel tracks, he bought two ex-army 4 wheel drive Marmon Herringtons that he converted to steel-bodied open vehicles for his transport run. One of the buses was known locally as 'the iron horse'. Because of the poor condition of the roads, when tides permitted, he ran his vehicles along the Amity-Point Lookout Beach. Ernie Odlem and Oliver Litherland ran bus and freight services in the township of Amity. Bonty Dickson ran a similar service at Dunwich. Dickson always dreamed of a trans-island road making Dunwich the point of entry for all Stradbroke. His determination and personal effort culminated in the construction of a bitumen road during his lifetime. In the initial stages of construction he induced locals to volunteer their time and machinery to clear the way. One steep, soft hill was known locally as 'Barry's Dive' after its maker (Barry Litherland), a volunteer from Amity. It was made trafficable by the broken concrete taken from the war-time air-raid shelters at Dunwich. The combination of broken concrete of uneven sizes and a steep drop gave every motorist a

Plate 96 MV 'Mirana' at the Amity jetty in the mid-1950s. In January 1943, Hayles' two river craft, 'Mirimar' and 'Mirabel', were requisitioned by Australian Army Water Transport. The boats transported troops and equipment to Moreton Bay garrisons, performed salvage work, and, on occasions, towed targets for coastal artillery practice. A civilian service for essential cargo and passengers for Moreton, Stradbroke, and Bribie Islands was maintained by the Army. (Joan Gregory)

thrill as he took 'the Dive'. The main Dunwich-Point Lookout road was named Dickson Way to recognise the perseverance of Bonty Dickson.

Island roads improved gradually, and in about 1957 the small vehicular ferry 'Kaboora' commenced a more or less regular service from Redland Bay to Dunwich. As there were no landing ramps on Stradbroke, skipper Jack Clark landed vehicles at high tide on the beach near the Dunwich Baths. This hazardous operation sometimes resulted in complete immersion of vehicles in saltwater.

In 1964, Stradbroke Ferries built the vehicular ferry 'Myora' for the run from Redland Bay to Dunwich. In the same year Vic Brook and Tony Durbidge built and operated 'The Lookout' from Paxton Street, Cleveland. This was the first of the drive-on and -off ferries. Around 1974 Stradbroke Ferries added 'Moongalba' to their fleet and moved its operations to the newly-dredged Toondah Harbour at Cleveland.

In 1973, Riverside Transport built 'Riverside Venture' which ran from Redland Bay. In 1977, in response to increasing demands for transport to the island, the barge was lengthened by 19.5 metres, the wheelhouse was raised 2 metres, and the 'Riverside Venture' could accommodate 44 cars.

Plate 97 The 'Kaboora', formerly the 'Amazon', an ex-army ALC 40, on the beach at Dunwich before ramps were constructed

Plate 98 'The men who commenced the east coast road' on 'Turk', their vehicle with a horn which sounded like a turkey. Standing: Allan Howe, Vince Tripcony, John Ramsden; seated Bonty Dickson, Doug Anning; inset Peter Hayles; absent George Doyle. Island roads were in a deplorable state when the area was placed under the control of Redland Shire Council in 1947. Bonty Dickson was Stradbroke's first representative in Council. He served until his resignation in 1968. (The late Bonty Dickson)

Plate 99 The 'Lookout', 1964, first of the drive-on and -off ferries to the island (Authors)

When the emphasis changed from launch to vehicular ferry transport, Hayles' launch services from Manly boat harbour and North Quay were phased out. Harold Walker operated his launch 'Vega' from Cleveland for daily commuters from Dunwich, and for years transported the island's secondary school children attending Cleveland High. He also discontinued his operation with the shift from launch to vehicular ferry transport.

In 1983 the Island Charter and Taxi Service commenced a fast water taxi connecting Cleveland and Dunwich. When the Groom family took over the operation three years later, the taxi became known as the Flyer. Stradbroke Ferries commenced a regular Cleveland-Dunwich water taxi service that connected with the Cleveland-Brisbane rail. This made it possible to reach the heart of the city in a relatively short time. In 2003, Sea Stradbroke put a second vehicular ferry on its Dunwich-Cleveland run.

Plate 100 MV 'Vega' at Dunwich jetty, early 1970s (Authors)

Plate 101 Stradbroke Ferries Vehicle Barge docked at Toondah Harbour, Cleveland preparing to load vehicles for Dunwich

Plate 102 'Karingal' at Amity Point preparing to load vehicles for Moreton Island

The Future

Plate 102 Many people and organisations demonstrated their opposition to a proposal to link North Stradbroke Island to the mainland. The 1984 anti-bridge march, Point Lookout (Authors)

North Stradbroke Island lies close to Brisbane and densely-populated southeast Queensland. Serviced by vehicular ferries and fast water taxis with a connecting train to the heart of Brisbane city, island residents can now commute daily to a mainland workplace. The island isolation that Stradbroke once experienced is no longer a major issue.

Everyone would accept that Stradbroke has changed dramatically in the last 20 years and that further development of the island will proceed at an increasing rate. This, however, is where the problems and conflicts begin. Defining what is 'best' for the island presents insurmountable difficulties because it means something different to everyone with a vested interest in the island's future.

The value of Stradbroke There are four ways of looking at the value of Stradbroke.'[113]

It can be viewed in *economic* terms – cold, hard cash – which can be quantified easily. Land values have skyrocketed. Stradbroke's minerals in 2003 were worth $61.8 million to the mining company and employees. The other industries – fishing, oystering, timber-getting, and tourism – all provide incomes which enable people to work on and to enjoy the island. Stradbroke has become an invaluable water source for Redland Shire's bay islands and the population expansion on the mainland; however, excessive water extraction could be to the detriment of the island's wetlands.

Stradbroke's *usefulness* to people is more difficult to quantify. The island's headlands, lakes, forest habitats, and bay shores are priceless resources for teachers and students of all ages; and its natural atmosphere and easygoing lifestyle attract many thousands of tourists each year. How can you place an economic value on a weekend spent in beautiful surroundings camping, fishing, walking or surfing?

The island has special *symbolic* value. The Aborigines have a cultural link with the island. Value can be attached with emotion, to a symbol or concept: Stradbroke as an island, a quiet refuge from the rush and bustle of Brisbane; Stradbroke unspoiled, a symbol of environmental concern for many people .

The island also has *intrinsic natural worth*. This derives from its special vegetation types and land forms, rare plants and animals, the special characteristics of its plant and animal communities which confer on it a unique quality. Intrinsic worth and symbolic value cannot be quantified. Are a koala or a rare flower or a jabiru worth anything?

As all these values are important in maintaining quality in our lives, it is reasonable to suggest that those who are in a position to control Stradbroke's future should ensure that all values are safeguarded.

Redland Shire Council is the local authority responsible for Stradbroke's future. Council's 1975 aims with regard to the island were to: 'protect areas of ecological significance, make the island readily accessible, provide adequate tourist facilities, secure adequate restoration of mined areas, provide inhabitants with facilities normally associated with normal living'.[14]

However, with the exception of 'making the island more accessible' and providing some infrastructure, unfortunately, in the main, Council's planning has not achieved the set goals.

Plans by the dray-load Yet the means to achieve these planning goals already exist. Stradbroke is probably the best understood and most planned for part of Queensland. Detailed information on its natural systems has been published in many scientific papers, and nine works covering aspects of the island's future have been prepared by experts: the Moreton Region Growth Strategy Investigation (1974), The Moreton Region Non-urban Land Suitability Study (1974), the Redland Shire Council Strategic Plan for North Stradbroke Island (1975), a plan to preserve habitats for mammals and birds (McEvoy, 1975), an analysis of the resources and management of the island (Rogers, 1975), an analysis of the future of the sandmining and tourism industries on the island (Fitzgibbons, 1975), and three preliminary statements by the Highway Planning Branch (1978-1979) on the proposal to build a bridge to the island and the impacts of such development.

In addition to Council's 1975 aims, the following documents provide a framework to ensure a sensitive approach to the planning of North Stradbroke Island: Redland Shire Open Space, Recreation, Conservation and Tourism Strategy (Wood, Thompson, Erickson, 1992); Status of Rare Plant Species of Myora Swamp, North Stradbroke Island (Bostock and Thomas, 1992); Environment Protection Act (1994); Coastal Protection and Management Act (1995); Development Control Plan 3, Point Lookout (1995); Integrated Planning Act (1997); Point Lookout Reserve R1781 (Stock, Olsen, Wilson, Brouwer, 1998); Water Reserve R884, North Stradbroke Island (Redland Shire Council, 1999); Redland Shire Council Environmental Inventory (1999); formulation of the Draft of the North Stradbroke Island/Minjerribah Planning and Management Study (2003).

Are these plans no more than wordy documents?

Plate 103 Brown Lake, one of the island's perched lakes, following heavy rain, 1974 (Authors)

Plate 104 Brown Lake following a dry spell during summer, 2001

Above: Plate 105 Canaipa Passage (Authors).

Opposite page above: Plate 106 Foredunes, Main Beach, before mining (Doug Hirst)

Opposite page below: Plate 107 Looking from the top of Mt Hargrave, 1995. All too frequent wild fires destroy wildlife and change plant regimes. (Jani Haenke)

'Wild' Stradbroke We acknowledge as a fact that there is probably no true wilderness in Australia, because humans have been everywhere and modified all places. Yet it is a fact also that Stradbroke is still in part a wild place, and that its wildness is a diminishing and irreplaceable resource. Should the low value placed on this resource by one section of the community (eg, developers, miners) always take precedence over the high value placed on it by others (eg, bird watchers, bush walkers, teachers)?

'Moreton Bay is the eastern gateway to Brisbane, and the northern approach by water to Southport and the Gold Coast. It is a jewel that must be preserved in a largely natural state, for future generations.'[24]

'North Stradbroke Island presents a classic instance of a comparatively small area, supporting an abundant fauna, being faced with potential development and a land use program (sandmining) which will destroy habitat and, ultimately, eliminate much of the fauna… If the fauna was sacrificed in the interests of development it would be a serious reflection on the way the present community managed the environment.'[69]

'The small area involved in wetlands and lakes makes preservation of these areas important: each hectare of

wetland is relatively more important than equivalent areas of high dune, partly because of the rarity of the land form but also because of its demonstrated complexity.'[89]

'Stradbroke Island is within easy reach of over 70 secondary schools and eight tertiary institutes in the Greater Brisbane Area. That Stradbroke is an island endows it with special characteristics in this increasingly populated region; an atmosphere conducive to discovery and exploration is created. Most of the basic problems of biology, ecology and the environment can be investigated in many different permutations in the area between Dunwich, Amity and Point Lookout. The area thus forms an ideal place for the development of a Field Studies Centre. The Dunwich Laboratories could be a nucleus for such a Centre – but to make the Field Studies Centre viable for posterity, suitable areas of the Island should be conserved as educational study areas. These areas should be under strict supervision but some collecting and experimentation should be allowed in order to demonstrate some ecological principles.'[97]

The price of progress Aborigines used the island's natural resources for thousands of years. It provided well for them. Europeans' impact of less than 200 years on the natural environment is not all to their credit.

Above: Plate 108 Gurri Nuggan, one of Tom Welsby's Aboriginal informants from Dunwich, about 1920 (Doug Anning)

Above left: Plate 109 Stradbroke is an ideal place for outdoor education, being close to Brisbane, but relatively unspoiled. (Authors)

Right: Plate 110 Grandfather Peter with boronias (Boronia safrolifera), Dunwich, 1960s (Doug Anning)

Environmental damage is often brushed aside as an inconvenient necessity of 'progress', but short-sightedness and selfishness also contribute substantially to unnecessary devastation.

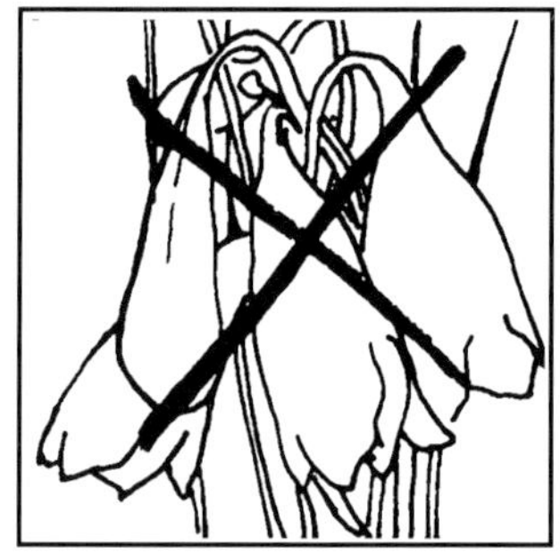

'Although the wholesale clearing of the vegetation for activities such as mining or agriculture are readily perceived as destructive of local ecosystems, these are not the only activities of man detrimental to the welfare of the vegetation. The indiscriminate collection of wild flowers is now frowned upon, yet this attitude has not always prevailed. About half-a-century ago Welsby described how great bunches of blossoms used to be collected for distribution in Dunwich or dispatched to Brisbane. It is probable that such intense harvesting led to the extinction of *Blandfordia grandiflora* (Christmas Bells) about Dunwich where, according to Scortechini (1882), it was formerly common in the swamps to the north of the town. Wholesale removal of blossom not only reduces the reproductive capacity of the plants, but it also results in the loss of fertility from the area, for pollen in particular is rich in nutrients.'[20]

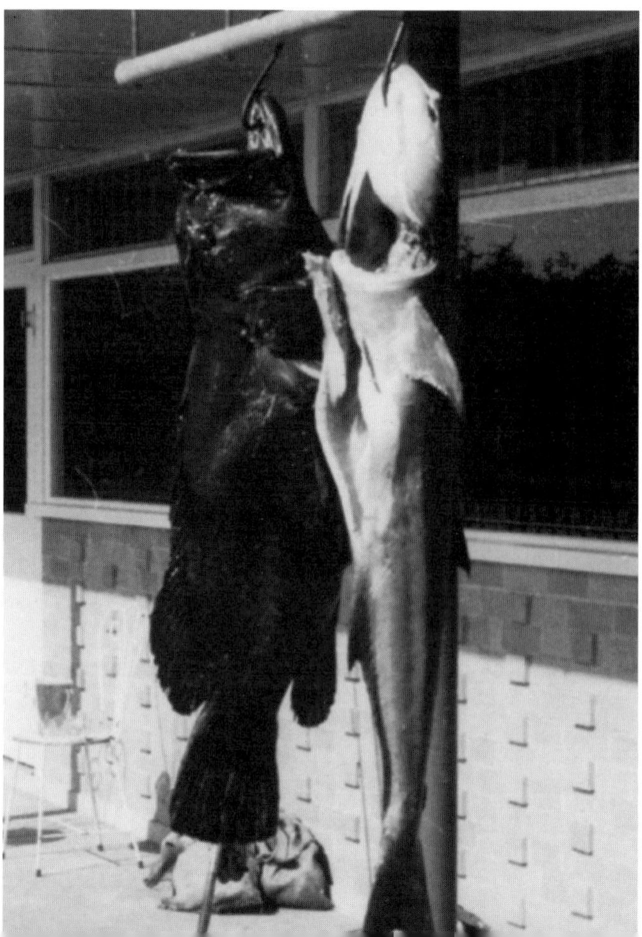

Plate 111 Black King and Cod caught near Point Lookout (Authors)

The yellow Everlasting Daisy (*Helichrysum bractetum*) once covered the exposed Point Lookout headlands. Through constant picking of flowers, the daisy no longer occurs on the headlands. Over-collection is also responsible for the disappearance from the Point Lookout rocky foreshore of *Melanerita melanotragus*, a tasty shellfish when boiled and picked from the shell with a pin.

Tom Welsby compared his fishing in 1931 to '...the old days of big snapper fishing, when the fish had been thick it was not uncommon for mackerel also to take a deep down bait. When two big snapper and one mackerel were on a three-hook line, the playing and landing of the trinity was deeply interesting.'[105]

Today, an all-day fishing trip to those same grounds might yield, with luck, just a couple of snapper.

Mangroves: 'Western man was attracted to tropical shores for a variety of reasons but certainly not to their mangroves. After all, it is difficult to see much of value in foetid, slimy mud, hordes of marauding insects, and a seemingly endless obstacle course of contorted and tangled vegetation, all thwarting access to and from dry land.

Plate 112 Ellie Durbidge and the late Kath Walker in the Grey Mangrove forest near Moongalba, c1980 (Authors)

'In the name of progress, therefore, and lacking better perception, it is not entirely surprising that thousands of hectares of these forests have been lost worldwide in programs of reclamation.'[13]

'What a splendid site this noble Bay would form for a Dutch Colony, *with the fee-simple granted to them of all the land they might reclaim from the waters!* It will be a magnificent field, by and by, for engineering. And gloomy and cheerless and apparently valueless as they are at present, even these mangrove forests will, ere long, form a mine of wealth to enterprising colonists; for their ashes form an excellent alkali for the manufacture of soap, and as tallow will in all likelihood be very cheap at Moreton Bay for a long time to come.'[58]

'...mangroves are not, for the most part, the prime cause of midge infestation. Where mangroves are essential to a midge or fortuitously assists its breeding, they do so in such a limited (possibly 1% or less) part of their whole extent that it is quite unreasonable to destroy large areas for pest control. Indeed, there are situations where deliberate planting of mangroves may be the best method of dealing with the problem.'[85]

Mangrove habitats on North Stradbroke island are seriously threatened. 'The recent discovery of such a rare mammal as the False Water Rat [Water Mouse] in these mangroves further emphasises the value of this unappreciated habitat.'[109]

'Overseas studies have shown that in some coastal waters 80% of fish caught commercially were linked to food chains dependent ultimately on mangroves.'[61]

Right: Plate 113 The Water Mouse (False Water Rat) (Xeromys myoides) is one of the rarest mammals in Australia. The population at Myora provided scientists from the Queensland Museum with the first opportunity to study the status and ecology of this poorly known species. (Queensland Museum)

Below: Plate 114 Roots of the Red Mangrove, near Blaksley Lagoon (Authors)

Introduced pests: 'Domestic dogs and cats forage near and far from home, obeying instincts to kill, without in fact depending for an existence on the animals they kill. They return home for a meal which may be minced kangaroo from a tin.'[2]

'The success of *Bufo marinus* (the Cane Toad) in establishing itself in Australia is attributable to its adaptability, resilience, high breeding potential, and relative freedom from predators...It has a deleterious effect on some native vertebrate fauna'.[26]

Right: Plate 115 Domestic cats and dogs abandoned or neglected by their owners and running wild are a threat to wildlife. (Queensland Museum)

Below: Plate 116 One native animal, the Freshwater Snake (or Common Keelback) can eat young Cane Toads, their tadpoles, and eggs. Even these are toxic to most other native animals. The Freshwater Snake cannot, however, cope with larger toads. Freshwater Snake killed while attempting to ingest a cane toad. (Queensland Museum)

Plate 117 Cane Toads are a curse to Stradbroke. They now plague all the townships on the island and, without doubt, have a deleterious effect on much of its fauna. In the 1960s there were only a few toads on the island, at Dunwich,apparently the progeny of deliberate introductions. Since that time they have infested all northern, disturbed environments on the island. Cane Toads affect native fauna both directly and indirectly. They are toxic to almost everything that feeds on them, so frog-eating native animals like some birds and snakes are especially vulnerable. They also affect native frog populations adversely by taking over their breeding grounds and eating their insect prey. New-born Death Adders feed solely on frogs (and toads that they mistake for frogs). The outcomes are invariably fatal. As a result, Death Adders are now uncommon where Cane Toads occur. (Queensland Museum)

Sandmining: The question of sandmining on Stradbroke is still a major issue. While the mineral sand-mining company has improved its record of stabilisation of the mined high dunes since a 1970 assessment, the company still has to answer for many destructive mining practices and disasters causing irreparable and permanent damage. Tailing spills and high dune slumps into the western swamp and tidal areas have occurred. In 1996, mining of the Gordon lease caused the flooding of three areas outside the lease boundary. During the 50-odd years of mining that Stradbroke has been subjected to, 'the intricate layers of sand that have been formed over thousands of years are replaced by a homogenous sand mass'.

Much of the island's landscape, perched water bodies, vegetation and fauna habitats have been destroyed or modified. The twin peaks of Mt Hutton were moved east of their original position. The mining company was responsible in 1991 for a 60,000-100,000 litre oil spill near Amity Swamp; the spill was not reported to the Department of Mines until three years later.

Plate 118 Lake Kounpee, a perched lake, in 1987. The mineral sandmining operation is visible on the dune behind the lake. (Authors)

'The reclamation and reafforestation carried out by the companies might not leave the mined areas exactly as nature had intended, but it would probably take a trained ecologist to pick the difference'.[118]

'...We consider that the ability of the mining companies to stabilise the defined areas of the mined high dunes on Stradbroke Island has not been demonstrated.'[114]

Situated above the island's groundwater table are perched lakes. These water bodies are destroyed when mining operations breach a lake's indurate layer. In November 1987, the dramatic drop in the water level of Lake Kounpee was the direct result of Consolidated Rutile mining too close to the lake. Since November 1987, the company has been endeavouring to restore the water level by pumping water back into the lake, pumping in mining files to seal the breach, and constructing betonite walls at the lake's perimeter. At one stage the lakebed was lined with betonite. This was disastrous, as it created a trap for wallabies that died lingering deaths in the betonite bog. Sixteen years later, the company has not restored Lake Kounpee.

Plate 119 Lake Kounpee, 2002. The mining company has been unsuccessful in attempts to restore the lake's water level. (Authors)

Upper left: Plate 120
Blaksley Lagoon, 1987, a perched water body drained by mining activities (Authors)

Lower left: Plate 121
The Blaksley Lagoon mining spill pushed the western swamp out into Moreton Bay in 1982. (Authors)

Opposite above: Plate 122
Mining operations, Blaksley Lagoon, 1980. Tailings cover parts of the freshwater swamp which fringes Stradbroke's bay shores. (Authors)

Opposite below: Plate 123
Steel water pipes for sandmining, south of Dunwich (Authors)

CONSERVATIONISTS COST JOBS

'Vast areas of North Stradbroke Island have been defaced by sandmining operations.'[24]

'...sand mining and residential usage [on Stradbroke] have high external costs. The way towards a socially optimal environment on Stradbroke Island would almost certainly be to ban sand mining, and to restrict the sale of residential blocks, together with the adoption of a plan for an adequate tourist environment'.[39]

(External costs are indirect costs not quantifiable in financial terms.)

The mining companies estimate that 12.5% of Stradbroke will be mined.[89] This sounds like a small amount of the island. If estimates are correct, nearly 88% of the island would remain undisturbed. Such figures can be misleading, however, giving a false sense of security about Stradbroke's future. The Eighteen Mile Swamp is currently under threat as mining operations move towards mining the high dune Enterprise lease adjacent to the swamp. Can we afford to lose it?

Water extraction: The island's giant sand masses 'soak up rain like giant sponges'. With water a valuable saleable commodity, Redland Shire Council has tapped into Stradbroke's groundwater resource to satisfy the growing Shire's demands.

'Groundwater occurs on Stradbroke as a lenticular body of fresh water. The groundwater mound is dynamic, involving deep recharge, continuous outflow in a seaward direction and discharge into coastal swamps and creeks. Stradbroke was first identified and considered as a potential source for mainland water supply in the early 1900s. However, active planning to "develop" this resource did not begin until the early 1980s when...[the] Council engaged consultants to investigate the supply of reticulated water for mainland consumers. It was decided that [the island] source was the "most feasible and cost-effective option" (John Wilson and Partners, 1984) for the long-term future water supply for the Shire. Natural Resources and Mines (then the Department of Natural Resources) did not require a formal environmental impact assessment of the Council proposal to export water off the island.'[62a]

Redland Water and Waste, an arm of the Council, began water mining in 1990. Natural Resources and Mines issues water licences. Redland, the largest user of island water, has a licence to extract up to 53ML/day.

Opposite above:Plate 124 Pumping water from swamps for mining lowers water tables. Peat is exposed at Yarraman, the head of the Eighteen Mile Swamp. (Authors)

Opposite below: Plate 125 Bund wall, Amity 1978. This wall was constructed illegally across Wallum Creek, outside a mining lease, to exclude saltwater and to provide freshwater for mining. It caused the destruction of a large tract of mangroves. The wall was removed followingsubmissions by concerned organisations and individuals. (Authors)

'The technical knowledge of the resource is critical to accurate definition, reliable forecasting and wise decision-making. Although there have been considerable data collection and analysis to assist decision-making, the management approach has been ad hoc and on a needs be basis. Moreover, there exist many gaps in the knowledge basis, uncertainties and unproven assumptions that have been consistently ignored over time.'[62a]

'A common side effect of mining, pumping fresh water from lagoons and swamps upsets water tables and vegetation communities.'[18]

Currently 70% of the Shire's water is drawn from Stradbroke Island. The Council anticipates 40,000 additional residents by 2012, yet it has made no provisions for increased water collection or for water recycling measures. Evidently, the plan is that the Shire's growing population will consume increasing amounts of island water, even though the island's resources are by no means assured of being unlimited.

Waste disposal: One of the major problems associated with Stradbroke's future is the disposal of waste – sewage, litter, obsolete machinery – generated by those who use the island. Redland Shire Council transports all household waste and old car bodies to the main-

Below: Plate 126 Some people use island lakes and swamps as rubbish tips. (Authors)

Opposite below: Plate 127 Street scene, Point Lookout, 1980s. This is no longer a familiar sight. (Authors)

land. Similarly, sewage solids are transported to the mainland, while filtered sewage effluent is still mostly disposed of in filtration ponds on the island.

'Control of litter-disposal will be imperative if the island's fragile ecosystems which have developed on extremely infertile soils are to be conserved; contamination with phosphatic nutrients can lead to marked changes in the ecosystems.'[97]

Run-off from septic tanks has for many years been finding its way down to headlands and beaches, causing problems of enriched nutrients that encourage weeds, and also creating unsightly concentrations of polluted slicks. Weeds infest many parts of the island, and there is yet to be an effective weed management program.

People are more ready to acknowledge the detrimental effects of plastics and other litter on marine life, and initiatives on the island to reduce the use of plastic bags have the potential to save many marine animals from choking deaths. Some enlightened coastal councils have begun to ban cigarette butts on beaches, another threat to turtles and fish that mistakenly consume these poisonous pellets.

Recent generations will go down in history as 'the greatest and most destructive vandals of all time'.[53]

Managing the coastal zone: Many plans covering aspects of North Stradbroke's future have been prepared, and the island is well-documented in scientific terms, but it is not a shining example of what can be done to maintain natural assets alongside development. There have been some monumental blunders despite both knowledge and plans. The Fraser Island Management Plan lists among its objectives provision of 'a range of recreational uses designed to optimise the island's recreational opportunities with minimum interference with the natural character of the island and to facilitate their use...'[55] It seems reasonable to suggest that this should also apply to North Stradbroke.

'It is part of the work of the National Parks and Wildlife Service to identify and reserve representative examples of each landscape type...'[92]

One National Park [Blue Lake-Tortoise Lagoon National Park] has been declared on the island but this

Below: Plate 128 Mangroves buried by mining tailings, Amity, 1980 (Authors)

Bottom: Plate 129 Filled foreshores, Amity, 1980. The local authority sacrificed mangrove forests and saltmarshes to provide camping facilities which could have been located elsewhere. (Authors)

is in a poor state. Recently an additional 67 ha have been added to this park giving it a total size of 512 ha.

'Management of the coastal zone of Australia is usually in the hands of local government authorities. Until recently these authorities have tended to use similar planning methods and criteria for all the land under their authority, without distinguishing the shoreline very much from the remainder, apart from some consciousness that waterfront land was more popular, and therefore more valuable for real estate development.

'Within the last decade it has come to be recognised that there are good reasons for considering that the management of shorelines (both of the ocean and of the estuaries) present special problems...Perhaps the real problem is that we do not, as a nation, have a national policy of coastline management, nor any body authorised to formulate or carry out such a policy.'[77]

Below: Plate 130 Camping ground, Point Lookout, Easter, 1980 (Authors).

Bottom: Plate 131 Four wheel drives have made the beach frontal dunes accessible to campers. This is destructive to beach fauna and dune vegetation. Flinders Beach, 1998 (Authors)

The bridge proposal: The proposal to link North Stradbroke and the mainland with a bridge was a controversial issue for many years. While it was argued that Stradbroke would undoubtedly develop more rapidly with a bridge, opponents have asked: at what cost to the island? Very few residents and visitors wanted to trade what the island has to offer for travel by highway to the island.[1] Residents, the boating fraternity, environmentalists and the general public demonstrated opposition to the bridge proposal in no uncertain terms.

In May 1988, the State Cabinet announced that bridge construction was not economically viable. However, some Russell Island residents who were promised a bridge by real estate agents still agitate for a bridge connection to their island. Such a proposal is a threat to Stradbroke. A bridge connection from Russell to Stradbroke would simply be a matter of bridging narrow Canaipa Passage.[45, 46, 47, 48, 49]

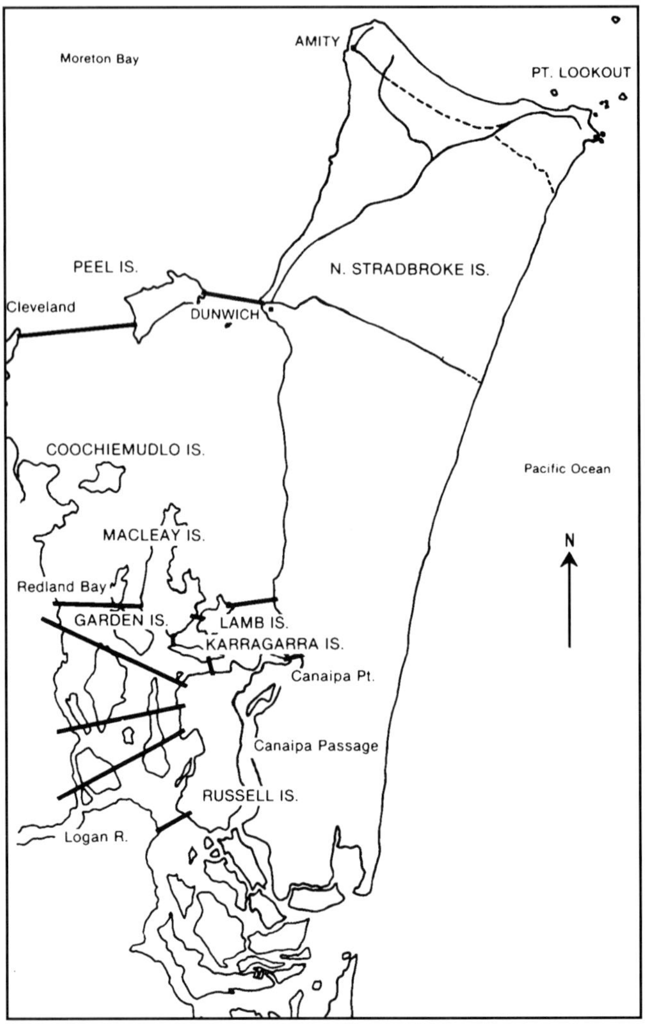

Fig 33 Possible mainland-island bridge/road links

Top: Plate 132 Before 4 wheel drives turned beaches into highways: migratory birds, Amity Bar, 1931. Under the International Ramsar Agreement to protect migratory birds, Stradbroke is a declared site. (Stubbins)

Bottom: Plate 133 Beaches are no longer places for birds and passive recreation. A beach permit from Redland Shire Council allows driving on beaches. Flinders Beach, 1998 (Authors)

Tourism: With sandmining operations expected to wind down as mineral deposits are exhausted, many people anticipate tourism replacing mining as a major feature of the local economy.

Stradbroke's superb coastal landscape, magnificent beaches, fishing and camping have always offered a low-key, family-oriented holiday. Board riders are attracted to the rolling surf. In winter, people come to witness the northern migration of the giant Humpback Whales. There is growing awareness of the island's colourful social and cultural history.

The appeal and value of Stradbroke as a premier tourism destination is not in question. The issue is how to manage the future development of a valuable, unique resource. Will this development exploit the island's natural attractions, or be an agent to help preserve them? Intense promotion by transport operators, Redland Shire Council, and real estate interests has created an explosion in the numbers of tourists visiting the island. With the rapid population increase in southeast Queensland, it is estimated that visitor numbers can rise to 400,000 per year.[119]

Will the island's uniqueness be destroyed under the sheer weight of numbers? Beaches that once were places for passive recreation have become highways for 4 wheel drives. Frontal dune camping is destroying dune vegetation of sea birds' resting and nesting sites. High-speed watercraft propellers are killing turtles and dugong.

Above: Plate 134 Four wheel drives make access to some areas easy, but they can be very destructive. (Authors)

Left: Plate 135 Blue Lake is in the island's only National Park. Visitors unwittingly damage its shores, 1980. (Authors)

'Isolation has meant that the island has remained a sleepy backwater, a patch of frontier 40 km from the city. This has attracted a group of tourists and settlers who appreciated the extensive natural forests, and the pleasant beaches. These assets, if retained, will undoubtedly form the basis for long term use of the area.'[89]

Above: Plate 136 While only 40 km from Brisbane, much of North Stradbroke Island remains relatively wild – but for how long?

As further land releases have been frozen by Native Title claims, the Council has been allowing multiple dwellings on undersized sites at Point Lookout, the island's major tourism destination. The increased population density erodes the character of the town, and causes problems associated with additional vehicles and noise nuisance. The result is loss of amenity for all.

Inevitably, Stradbroke is the subject of a clash of values as the competing visions for the island's future are played out. Is Stradbroke's fate to be developed as a clone of the rest of the coast, or can it retain the peaceful, natural character that has attracted visitors for generations?

Plate 137 Access to Main Beach in the good old days (Joan Gregory)

> *'Straddie is a truly beautiful place…The black swans, the kingfishers, the tawny frogmouths: amazing bird life. The inland lakes, and the goannas lazing on walking tracks. The meandering foreshore at Dunwich, its underwater world exposed at low tide. The springs at Myora: the cockle and oyster shell middens on their banks, freshwater crays nibbling at your toes. The majestic Point Lookout: dolphins surfing and whales breaching. Dugongs grazing on sea-bed grass.*
>
> *'And the surf, the beaches, the dunes, the breezes, the air – your island's natural splendour!*
>
> *'But there's a feeling, too, about your island, and it comes over me as soon as the barge leaves the mainland. The feeling that I'm leaving behind a frenetic, city buzz and retreating into a quieter, calmer, more reflective zone.*
>
> *'People who love to stay on the island crave this feeling as their holidays approach and mourn its loss as they return to the mainland; people who live on the island have this sense etched in their being, and fondly call it "Straddie-time".'*

In August 2004, Dunwich State School celebrated its centenary. Above is an excerpt from the speech by the Governor, Ms Quentin Bryce, to islanders on the occasion.

MINJERRIBA

Minjerriba[1] was a giant in the sun
His green back coated with cypress and gum.
Belly bloated with rich grains of sand
Eyes brimming with waters so cool,
He stretched for miles in the sun.
And Pacific on the east
Quandamooka[2] on the west
Bathed this giant in the sun.

But Minjerriba's back is now broken;
Men came and tore out his guts;
Stole his rich grains of sand,
Stripped his cloak of cypress and gum,
Drained water from his ageless eyes
And weakened this giant in the sun.

Oh man! With your machinery and science,
Your greed and callous disregard,
When your savage looting and lying is done,
Will the Gods in the future
If future there is
Spare you your place in the sun?

Oodgeroo Noonukul (Kath Walker)

Minjerriba[1] and Quandamooka[2]
are the Aboriginal names for
Stradbroke Island and
Moreton Bay respectively.

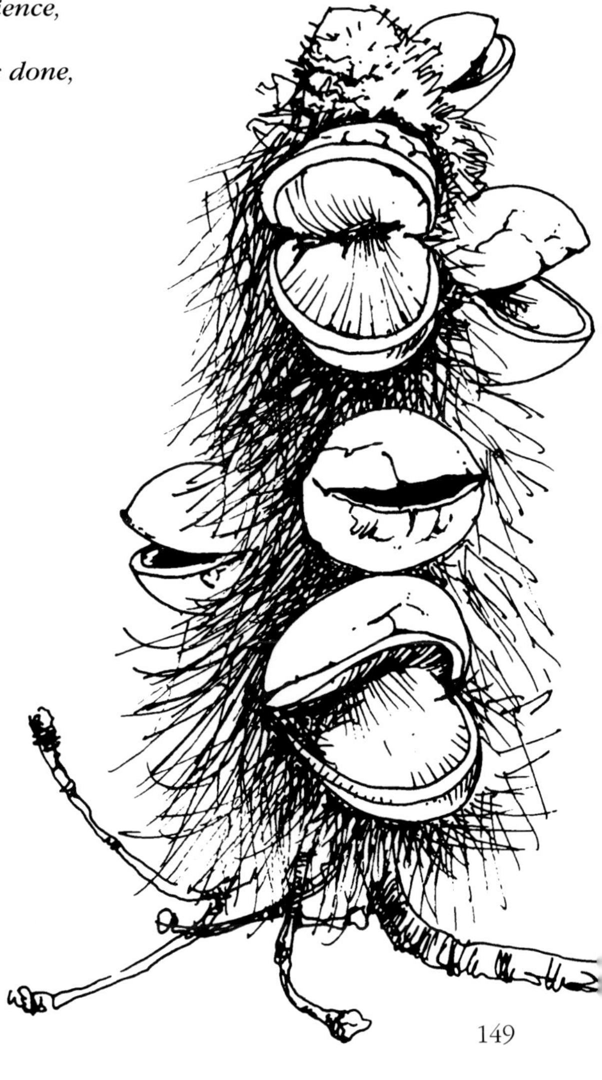

APPENDIX I

Fauna of Stradbroke

Mammals

Short-beaked Echidna
Tachyglossus aculeatus
Common Planigale
Planigale maculata
Northern Brown Bandicoot
Isoodon macrourus
Koala
Phascolarctos cinereus
Squirrel Glider
Petaurus norfolcensis
Feathertail Glider
Acrobates pygmaeus
Agile Wallaby
Macropus agilis
Eastern Grey Kangaroo
Macropus giganteus
Swamp Wallaby
Wallabia bicolor
Black Flying-fox
Pteropus alecto
Grey-headed Flying-fox
Pteropus poliocephalus
Little Red Flying-fox
Pteropus scapulatus
Common Blossom-bat
Syconycteris australis
Beccari's Freetail-bat
Mormopterus beccarii
Southern Freetail-bat
Mormopterus planiceps
White-striped Freetail-bat
Nyctinomus australis
Gould's Wattled Bat
Chalinolobus gouldii
Hoary Wattled Bat
Chalinolobus nigrogriseus
Little Broad-nosed Bat
Scotorepens greyii
Eastern Broad-nosed Bat
Scotorepens orion
Water Rat
Hydromys chrysogaster
Grassland Melomys
Melomys burtoni
Fawn-footed Melomys
Melomys cervinipes
House Mouse
Mus musculus
Swamp Rat
Rattus lutreolus
Black Rat
Rattus rattus
Pale Field-rat
Rattus tunneyi
Water Mouse (formerly called False Water-rat)
Xeromys myoides
Leopard Seal
Hydrurga leptonyx
Cat
Felis catus
Red Fox
Vulpes vulpes
Horse
Equus caballus
Dugong
Dugong dugon
Southern Right Whale
Eubalaena australis
Bryde's Whale
Balaenoptera edeni
Humpback Whale
Megaptera novaeangliae
Blainville's Beaked Whale
Mesoplodon densirostris
Sperm Whale
Physeter macrocephalus
Long-finned Pilot Whale
Globicephala melas
Risso's Dolphin
Grampus griseus
Killer Whale
Orcinus orca
Melon-headed Whale
Peponocephala electra
Indo-Pacific Hump-backed Dolphin
Sousa chinensis
Striped Dolphin
Stenella coeruleoalba
Long-beaked Bottlenose Dolphin
Tursiops aduncus
Bottlenose Dolphin
Tursiops truncatus

Amphibians

Tusked Frog
Adelotus brevis
Clicking Froglet
Crinia signifera
Wallum Froglet
Crinia tinnula
Brown-striped Marshfrog
Limnodynastes peronii
Spotted Marshfrog
Limnodynastes tasmaniensis
Scarlet-sided Pobblebonk
Limnodynastes terraereginae
Great Brown Broodfrog
Pseudophryne major
Sandy Gungan
Uperoleia fusca
Green Treefrog
Litoria caerulea
Cooloola Sedgefrog
Litoria cooloolensis
Eastern Sedgefrog
Litoria fallax
Wallum Rocketfrog
Litoria freycineti
Graceful Treefrog
Litoria gracilenta
Striped Rocketfrog
Litoria nasuta
Wallum Sedgefrog
Litoria olongburensis
Emerald-spotted Treefrog
Litoria peronii
Cane Toad
Bufo marinus

Reptiles

Loggerhead Turtle
Caretta caretta
Green Turtle
Chelonia mydas
Leatherback Turtle
Dermochelys coriacea
Eastern Snake-necked Turtle
Chelodina longicollis

House Gecko
Hemidactylus frenatus
Lesueur's Velvet Gecko
Oedura lesueurii
Burton's Legless Lizard
Lialis burtonis
Diporiphora australis
Bearded Dragon
Pogona barbata
Sand Monitor
Varanus gouldii
Lace Monitor
Varanus varius
Anomalopus Verreauxii
Calyptotis scutirostrum
Carlia vivax
Cryptoblepharus virgatus
Ctenotus arcanus
Ctenotus robustus
Copper-tailed Skink
Ctenotus taeniolatus
Major Skink
Egernia frerei
Eulamprus martini
Eastern Water Skink
Eulamprus quoyii
Eulamprus tenuis
Lampropholis delicata
Lampropholis guichenoti
Fire-tailed Skink
Morethia taeniopleura
Ophioscincus truncatus
Saiphos equalis
Saproscincus oriarus
Eastern Blue-tongued Lizard
Tiliqua scincoides
Carpet Python
Morelia spilota
Brown Tree Snake
Boiga irregularis
Common Tree Snake
Dendrelaphis punctulata
Freshwater Snake
Tropidonophis mairii
Death Adder
Acanthophis antarcticus
White-crowned Snake
Cacophis harriettae
Yellow-faced Whip Snake
Demansia psammophis
Black Whip Snake
Demansia vestigiata
Black-bellied Swamp Snake
Hemiaspis signata
Stephens' Banded Snake
Hoplocephalus stephensii
Eastern Tiger Snake
Notechis scutatus
Red-bellied Black Snake
Pseudechis porphyriacus
Eastern Brown Snake
Pseudonaja textilis
Eastern Small-eyed Snake
Rhinoplocephalus nigrescens
Rough-scaled Snake
Tropidechis carinatus
Bandy-bandy
Vermicella annulata
Acalyptophis peronii
Olive Sea Snake
Aipysurus laevis
Disteira kingii
Disteira major
Hydrophis elegans
Yellow-bellied Sea Snake
Pelamis platurus

Birds

Australian Brush-turkey
Alectura lathami
Brown Quail
Coturnix ypsilophora
King Quail
Coturnix chinensis
Wandering Whistling-duck
Dendrocygna arcuata
Musk Duck
Biziura lobata
Black Swan
Cygnus atratus
Australian Wood Duck
Chenonetta jubata
Pacific Black Duck
Anas superciliosa
Grey Teal
Anas gracilis
Chestnut Teal
Anas castanea
Hardhead
Aythya australis
Australasian Grebe
Tachybaptus novaehollandiae
Little Penguin
Eudyptula minor
Common Diving-petrel
Pelecanoides urinatrix
Southern Giant-petrel
Macronectes giganteus
Northern Giant-petrel
Macronectes halli
Cape Petrel
Daption capense
Tahiti Petrel
Pseudobulweria rostrata
Great-winged Petrel
Pterodroma macroptera
White-headed Petrel
Pterodroma lessonii
Providence Petrel
Pterodroma solandri
Kermadec Petrel
Pterodroma neglecta
Mottled Petrel
Pterodroma inexpectata
White-necked Petrel
Pterodroma cervicalis
Black-winged Petrel
Pterodroma nigripennis
Cook's Petrel
Pterodroma cookii
Gould's Petrel
Pterodroma leucoptera
Blue Petrel
Halobaena caerulea
Broad-billed Prion
Pachyptila vittata
Salvin's Prion
Pachyptila salvini
Antarctic Prion
Pachyptila desolata
Slender-billed Prion
Pachyptila belcheri
Fairy Prion
Pachyptila turtur

Streaked Shearwater
Calonectris leucomelas
Wedge-tailed Shearwater
Puffinus pacificus
Buller's Shearwater
Puffinus bulleri
Flesh-footed Shearwater
Puffinus carneipes
Sooty Shearwater
Puffinus griseus
Short-tailed Shearwater
Puffinus tenuirostris
Fluttering Shearwater
Puffinus gavia
Hutton's Shearwater
Puffinus huttoni
Little Shearwater
Puffinus assimilis
Wandering Albatross
Diomedea exulans
Black-browed Albatross
Diomedea melanophrys
Shy Albatross
Diomedea cauta
Grey-headed Albatross
Diomedea chrysostoma
Yellow-nosed Albatross
Diomedea chlororhynchos
Buller's Albatross
Diomedea bulleri
Sooty Albatross
Phoebetria fusca
Light-mantled Sooty Albatross
Phoebetria palpebrata
Wilson's Storm-petrel
Oceanites oceanicus
White-faced Storm-petrel
Pelagodroma marina
Black-bellied Storm-petrel
Fregetta tropica
White-bellied Storm-petrel
Fregetta grallaria
Red-tailed Tropicbird
Phaethon rubricauda
White-tailed Tropicbird
Phaethon lepturus
Australasian Gannet
Morus serrator
Masked Booby
Sula dactylatra
Red-footed Booby
Sula sula
Brown Booby
Sula leucogaster
Darter
Anhinga melanogaster
Little Pied Cormorant
Phalacrocorax melanoleucos
Pied Cormorant
Phalacrocorax varius
Little Black Cormorant
Phalacrocorax sulcirostris
Great Cormorant
Phalacrocorax carbo
Australian Pelican
Pelecanus conspicillatus
Great Frigatebird
Fregata minor
Lesser Frigatebird
Fregata ariel
White-faced Heron
Egretta novaehollandiae
Little Egret
Egretta garzetta
Eastern Reef Egret
Egretta sacra
White-necked Heron
Ardea pacifica
Great Egret
Ardea alba
Intermediate Egret
Ardea intermedia
Striated Heron
Butorides striatus
Nankeen Night Heron
Nycticorax caledonicus
Little Bittern
Ixobrychus minutus
Black Bittern
Ixobrychus flavicollis
Australian White Ibis
Threskiornis molucca
Straw-necked Ibis
Threskiornis spinicollis
Royal Spoonbill
Platalea regia
Black-necked Stork
Ephippiorhynchus asiaticus
Osprey
Pandion haliaetus
Black-shouldered Kite
Elanus axillaris
Letter-winged Kite
Elanus scriptus
Whistling Kite
Haliastur sphenurus
Brahminy Kite
Haliastur indus
White-bellied Sea-eagle
Haliaeetus leucogaster
Swamp Harrier
Circus approximans
Brown Goshawk
Accipiter fasciatus
Collared Sparrowhawk
Accipiter cirrhocephalus
Wedge-tailed Eagle
Aquila audax
Little Eagle
Hieraaetus morphnoides
Brown Falcon
Falco berigora
Australian Hobby
Falco longipennis
Peregrine Falcon
Falco peregrinus
Nankeen Kestrel
Falco cenchroides
Brolga
Grus rubicunda
Buff-banded Rail
Gallirallus philippensis
Lewin's Rail
Rallus pectoralis
Bush-hen
Amaurornis olivaceus
Baillon's Crake
Porzana pusilla
Spotless Crake
Porzana tabuensis
Purple Swamphen
Porphyrio porphyrio
Dusky Moorhen
Gallinula tenebrosa

Painted Button-quail
Turnix varia
Latham's Snipe
Gallinago hardwickii
Black-tailed Godwit
Limosa limosa
Bar-tailed Godwit
Limosa lapponica
Little Curlew
Numenius minutus
Whimbrel
Numenius phaeopus
Eastern Curlew
Numenius madagascariensis
Common Greenshank
Tringa nebularia
Terek Sandpiper
Xenus cinereus
Common Sandpiper
Actitis hypoleucos
Grey-tailed Tattler
Heteroscelus brevipes
Wandering Tattler
Heteroscelus incanus
Ruddy Turnstone
Arenaria interpres
Great Knot
Calidris tenuirostris
Red Knot
Calidris canutus
Sanderling
Calidris alba
Red-necked Stint
Calidris ruficollis
Sharp-tailed Sandpiper
Calidris acuminata
Curlew Sandpiper
Calidris ferruginea
Painted Snipe
Rostratula benghalensis
Comb-crested Jacana
Irediparra gallinacea
Bush Stone-curlew
Burhinus grallarius
Beach Stone-curlew
Esacus neglectus
Pied Oystercatcher
Haematopus longirostris
Sooty Oystercatcher
Haematopus fuliginosus
Black-winged Stilt
Himantopus himantopus
Pacific Golden Plover
Pluvialis fulva
Grey Plover
Pluvialis squatarola
Double-banded Plover
Charadrius bicinctus
Lesser Sand Plover
Charadrius mongolus
Greater Sand Plover
Charadrius leschenaultii
Black-fronted Dotterel
Elseyornis melanops
Masked Lapwing
Vanellus miles
South Polar Skua
Catharacta maccormicki
Great Skua
Catharacta skua
Pomarine Jaeger
Stercorarius pomarinus
Arctic Jaeger
Stercorarius parasiticus
Long-tailed Jaeger
Stercorarius longicaudus
Silver Gull
Larus novaehollandiae
Gull-billed Tern
Sterna nilotica
Caspian Tern
Sterna caspia
Lesser Crested Tern
Sterna bengalensis
Crested Tern
Sterna bergii
Roseate Tern
Sterna dougallii
White-fronted Tern
Sterna striata
Common Tern
Sterna hirundo
Arctic Tern
Sterna paradisaea
Little Tern
Sterna albifrons
Bridled Tern
Sterna anaethetus
Sooty Tern
Sterna fuscata
Whiskered Tern
Chlidonias hybridus
White-winged Black Tern
Chlidonias leucopterus
Common Noddy
Anous stolidus
Black Noddy
Anous minutus
Grey Ternlet
Procelsterna cerulea
White Tern
Gygis alba
Rock Dove
Columba livia
White-headed Pigeon
Columba leucomela
Spotted Turtle-dove
Streptopelia chinensis
Brown Cuckoo-dove
Macropygia amboinensis
Emerald Dove
Chalcophaps indica
Common Bronzewing
Phaps chalcoptera
Crested Pigeon
Ocyphaps lophotes
Diamond Dove
Geopelia cuneata
Peaceful Dove
Geopelia striata
Bar-shouldered Dove
Geopelia humeralis
Wonga Pigeon
Leucosarcia melanoleuca
Wompoo Fruit-dove
Ptilinopus magnificus
Superb Fruit-dove
Ptilinopus superbus
Rose-crowned Fruit-dove
Ptilinopus regina
Topknot Pigeon
Lopholaimus antarcticus
Glossy Black-cockatoo
Calyptorhynchus lathami

Galah
Cacatua roseicapilla
Sulphur-crested Cockatoo
Cacatua galerita
Rainbow Lorikeet
Trichoglossus haematodus
Scaly-breasted Lorikeet
Trichoglossus chlorolepidotus
Little Lorikeet
Glossopsitta pusilla
Australian King-parrot
Alisterus scapularis
Crimson Rosella
Platycercus elegans
Pale-headed Rosella
Platycercus adscitus
Swift Parrot
Lathamus discolor
Red-rumped Parrot
Psephotus haematonotus
Pallid Cuckoo
Cuculus pallidus
Brush Cuckoo
Cacomantis variolosus
Fan-tailed Cuckoo
Cacomantis flabelliformis
Horsfield's Bronze-cuckoo
Chrysococcyx basalis
Shining Bronze-cuckoo
Chrysococcyx lucidus
Little Bronze-cuckoo
Chrysococcyx minutillus
Common Koel
Eudynamys scolopacea
Channel-billed Cuckoo
Scythrops novaehollandiae
Pheasant Coucal
Centropus phasianinus
Powerful Owl
Ninox strenua
Southern Boobook
Ninox novaeseelandiae
Barn Owl
Tyto alba
Tawny Frogmouth
Podargus strigoides
White-throated Nightjar
Eurostopodus mystacalis
Australian Owlet-nightjar
Aegotheles cristatus
White-rumped Swiftlet
Collocalia spodiopygius
White-throated Needletail
Hirundapus caudacutus
Fork-tailed Swift
Apus pacificus
Azure Kingfisher
Alcedo azurea
Laughing Kookaburra
Dacelo novaeguineae
Forest Kingfisher
Todiramphus macleayii
Red-backed Kingfisher
Todiramphus pyrrhopygia
Sacred Kingfisher
Todiramphus sanctus
Collared Kingfisher
Todiramphus chloris
Rainbow Bee-eater
Merops ornatus
Dollarbird
Eurystomus orientalis
Noisy Pitta
Pitta versicolor
White-throated Treecreeper
Cormobates leucophaeus leucophaeus
Superb Fairy-wren
Malurus cyaneus
Red-backed Fairy-wren
Malurus melanocephalus
Spotted Pardalote
Pardalotus punctatus
Striated Pardalote
Pardalotus striatus
Weebill
Smicrornis brevirostris
Mangrove Gerygone
Gerygone levigaster
White-throated Gerygone
Gerygone olivacea
Brown Thornbill
Acanthiza pusilla
Red Wattlebird
Anthochaera carunculata
Little Wattlebird
Anthochaera chrysoptera
Striped Honeyeater
Plectorhyncha lanceolata
Noisy Friarbird
Philemon corniculatus
Little Friarbird
Philemon citreogularis
Blue-faced Honeyeater
Entomyzon cyanotis
Noisy Miner
Manorina melanocephala
Lewin's Honeyeater
Meliphaga lewinii
Yellow-faced Honeyeater
Lichenostomus chrysops
Mangrove Honeyeater
Lichenostomus fasciogularis
Fuscous Honeyeater
Lichenostomus fuscus
White-throated Honeyeater
Melithreptus albogularis
White-naped Honeyeater
Melithreptus lunatus
Brown Honeyeater
Lichmera indistincta
White-cheeked Honeyeater
Phylidonyris nigra
Eastern Spinebill
Acanthorhynchus tenuirostris
Scarlet Honeyeater
Myzomela sanguinolenta
Crimson Chat
Epthianura tricolor
Flame Robin
Petroica phoenicea
Rose Robin
Petroica rosea
Hooded Robin
Melanodryas cucullata
Pale-yellow Robin
Tregellasia capito
Eastern Yellow Robin
Eopsaltria australis
Grey-crowned Babbler
Pomatostomus temporalis
Varied Sittella
Daphoenositta chrysoptera

Golden Whistler
Pachycephala pectoralis
Rufous Whistler
Pachycephala rufiventris
Little Shrike-thrush
Colluricincla megarhyncha
Grey Shrike-thrush
Colluricincla harmonica
Black-faced Monarch
Monarcha melanopsis
Spectacled Monarch
Monarcha trivirgatus
White-eared Monarch
Monarcha leucotis
Leaden Flycatcher
Myiagra rubecula
Satin Flycatcher
Myiagra cyanoleuca
Shining Flycatcher
Myiagra alecto
Restless Flycatcher
Myiagra inquieta
Magpie-lark
Grallina cyanoleuca
Rufous Fantail
Rhipidura rufifrons
Grey Fantail
Rhipidura fuliginosa
Willie Wagtail
Rhipidura leucophrys
Spangled Drongo
Dicrurus bracteatus
Black-faced Cuckoo-shrike
Coracina novaehollandiae
Barred Cuckoo-shrike
Coracina lineata
White-bellied Cuckoo-shrike
Coracina papuensis
Cicadabird
Coracina tenuirostris
White-winged Triller
Lalage sueurii
Varied Triller
Lalage leucomela
Olive-backed Oriole
Oriolus sagittatus
Figbird
Sphecotheres viridis
White-breasted Woodswallow
Artamus leucorynchus
Dusky Woodswallow
Artamus cyanopterus
Grey Butcherbird
Cracticus torquatus
Pied Butcherbird
Cracticus nigrogularis
Australian Magpie
Gymnorhina tibicen
Pied Currawong
Strepera graculina
Torresian Crow
Corvus orru
Green Catbird
Ailuroedus crassirostris
Regent Bowerbird
Sericulus chrysocephalus
Satin Bowerbird
Ptilonorhynchus violaceus
Spotted Bowerbird
Chlamydera maculata
Richard's Pipit
Anthus novaeseelandiae
House Sparrow
Passer domesticus
Double-barred Finch
Taeniopygia bichenovii
Red-browed Finch
Neochmia temporalis
Nutmeg Mannikin
Lonchura punctulata
Chestnut-breasted Mannikin
Lonchura castaneothorax
European Goldfinch
Carduelis carduelis
Mistletoebird
Dicaeum hirundinaceum
Welcome Swallow
Hirundo neoxena
Tree Martin
Hirundo nigricans
Fairy Martin
Hirundo ariel
Clamorous Reed-warbler
Acrocephalus stentoreus
Tawny Grassbird
Megalurus timoriensis
Little Grassbird
Megalurus gramineus
Golden-headed Cisticola
Cisticola exilis
Silvereye
Zosterops lateralis
Russet-tailed Thrush
Zoothera heinei
Common Starling
Sturnus vulgaris
Common Myna
Acridotheres tristis

APPENDIX II: Street names and their origins

Point Lookout

Bambara Street: from the Stradbroke dialects, Coast Banksia (*Banksia integrifolia*)
Bigoon Street: from the Stradbroke dialects, jew fish
Billa Street: from the Stradbroke dialects, Bull Oak (*Casuarina glauca*)
Bimba Street: from the Stradbroke dialects, snapper fish
Booran Street: from the Stradbroke dialects, south wind
Boreen Street: from the Stradbroke dialects, whiting fish
Cutter Street: See Hopewell
Endeavour Street: Lieutenant James Cook, sailing along the eastern Australian coast in the 'Endeavour' in 1770, named Point Lookout
Galeen Street: from the Stradbroke dialects, cloud
Gindarra Street: from the Stradbroke dialects, the hills of Moreton Island
Hopewell Street: When the 'Porpoise' and 'Cato' were wrecked on Wreck Reef in 1803, Flinders, with the captain of the 'Cato' and 12 men, in one of the ship's cutters which the men named 'Hope', rowed over 700 miles to Sydney for help. To replenish the 'Hope's' water supply, Flinders landed four men in the vicinity of what is now known as Cylinder Beach
Midjimberry Road: from the Stradbroke dialects. Midyims (*Austromyrtus dulcis*) were berries eaten by the Aborigines
Mintee Street: from the Stradbroke dialects, Wallum Banksia (*Banksia integrifolia*)
Mooloomba Road: from the Stradbroke dialects, the north end of Stradbroke
Moongalba Road: the Aboriginal Mission at Myora
Mulung Street: from the Stradbroke dialects, bream fish
North Lane: William North ran cattle on North Stradbroke Island to supply the Dunwich Institution with beef. He ended his days as an inmate at Dunwich.
Prosperity Street: The ship 'Prosperity' in 1902 was wrecked just south of Point Lookout.
Timbin Road: from the Stradbroke dialects, north wind
Yarrong Road: from the Stradbroke dialects, sand

Amity Point

Ballow Street: Dr Ballow, Brisbane's Medical Officer, contracted typhoid fever and died in 1850 while attending typhus patients quarantined at Dunwich. Ballow Chambers in Brisbane is also named after the doctor. Dr Ballow was buried in the Dunwich Cemetery.
Bellerophon Street: named after one of the ships with which Matthew Flinders was associated
Berkeley Street: Mr E Berkeley was one of the passengers lost when the 'Sovereign' was wrecked in 1847, on the South Passage Bar.
Birch Street: Gustavus Birch, an Amity resident, 1850-1888, recorded island life and events in his 1873 diary.
Chettle Street: Mrs Chettle was one of the passengers lost when the 'Sovereign' was wrecked in 1847, on the South Passage Bar
Cook Street: named after the navigator, James Cook
Cumberland Street: named after one of the ships with which Matthew Flinders was associated
Dugong Street: see Gonzales
Ellis Street: The Hon John Ellis was an Amity resident known to Tom Welsby
Fernandez Street: see Gonzales
Flinders Road: after the navigator Matthew Flinders who, in the sloop 'Norfolk', surveyed northern Moreton Bay in 1799
Gonzales Street: Fernandez Gonzales, after being tried before the Sydney Court for disobeying orders at sea, and escaping severe punishment, came to Moreton Bay. He joined the crew of the pilot boat at Bulwer. Later he worked for Andrew Petrie gathering coral for lime burning. Finally he established himself as a dugong fisherman at Amity.
Gore Street: Mr and Mrs Robert Gore and their two children were lost when the 'Sovereign' was wrecked in 1847, on the South Passage Bar. The bodies of Mrs Gore and one of her children were washed up on the south end of Moreton Island
Kindara Street: When Welsby writes of Kindara, she lived at Moongalba. She was described as a well-known character 'even in the city as well as Moreton Bay'.
Miles Street: This was probably named after Canon Miles, an Anglican minister whose parish covered the islands of Moreton Bay. Undaunted by bad weather or transport difficulties, he was always on hand when needed. He was active until his death at the age of 87 in 1964.

Mirimar Street: Hayles Brothers built a kiosk at Amity Point in 1935 and commenced a launch service from North Quay, Brisbane. MV 'Mirimar', of 96 feet, was the first launch on the Stradbroke run.
Nuggin Street: Welsby tells the story of Tommy Nuggin who fell in love with Sarah. While Tommy was away serving with the Native Police Force, Adam, a 'bad man' of the Nigi Nigi of Bribie, stole Sarah. After Sarah's rescue by Tommy, Adam was never again sighted.
Porpoise Street: named after one of the ships with which Matthew Flinders was associated
Providence Street: named after one of the ships with which Matthew Flinders was associated
Sovereign Road: When the paddle steamer 'Sovereign' was wrecked on the South Passage Bar, on March llth 1847, 44 lives were lost.
Toompany Street: Toompany was one of the Aborigines who swam out in the surf to rescue ten people from the wrecked 'Sovereign'.

Dunwich

Ballow Road: see Dr Ballow listed in the streets of Amity
Barton Street: It seems reasonable to assume that this street was named after Dr F J Barton who became resident surgeon for Brisbane in 1850, when Dr D K Ballow died tending typhus patients at Dunwich.
Bingle Road: Captain John Bingle, sent in the sloop 'Sally', in 1822, 'to search for a large river supposed to exist between port Macquarie and Sandy Cape', surveyed some of Moreton Bay, but did not find the Brisbane River.
Cunningham Street: Allan Cunningham, a botanist and explorer, was sent with Oxley to Moreton Bay to help select a site for the new settlement and to collect specimens. On a later trip to Moreton Bay in 1828 he visited the newly formed establishment at Dunwich.
Finnegan Street: Finnegan was one of the three cast-away cedar-getters, who, wrecked on Moreton Island, crossed the South Passage in a canoe to Amity where they lived with the Aborigines for some months.
Flinders Avenue: see Norfolk
Fraser Street: The botanist Charles Fraser in 1828 accompanied Allan Cunningham on a trip exploring and collecting botanical specimens from the Moreton Bay region for the Royal Gardens of Kew.

Guy Crescent: Guy was one of the surveyors to survey the town of Dunwich.
Jack Lane: Captain Jack was one of the captains of the government steamer 'Otter' which serviced the Dunwich Institution.
Junner Street: Captain Junner was one of the captains of the government steamer 'Otter' which serviced the Dunwich Institution.
Lang Lane: Dr J Dunmore Lang (1799-1878) was a Scottish Presbyterian minister who worked tirelessly to establish schools in the Colony of New South Wales, and also to induce skilled British workers to settle in the colony.
Logan Lane: Captain Logan, Commandant of the Moreton Bay settlement, was ordered by Governor Darling to establish a depot at Dunwich. '1.5.28 Logan reported that the magazine at Dunwich was roofed and nearly completed, and the military and prisoners' barracks at Dunwich were finished and occupied.'
Mallon Street: The ship 'Emigrant' in 1850, arriving at Moreton Bay with 64 cases of typhus, was quarantined at Dunwich. When the ship's surgeon Dr George Mitchell died of typhus, Dr P W Mallon was sent from Brisbane but he also became ill. Dr Ballow, Brisbane's Resident Surgeon, went to Dunwich to help. Unfortunately he also contracted the fever and died. Dr Mallon recovered.
Mermaid Street: see Oxley. Major Lockyer also used the 'Mermaid' in 1825 to visit Moreton Bay to survey the Brisbane River.
Mitchell Crescent: see Mallon
Norfolk Street: Matthew Flinders, one of the world's great navigators, mapped most of the Australian coastline. In the sloop 'Norfolk' he charted northern Moreton Bay in 1799.
Otter Street: 'The Dunwich Institution was supplied from Brisbane by the Government Steamer "Otter", a steel, twin screwed vessel of 135 feet, her gross tonnage of 222 and her engine of 96 h p. She was built by Ramage and Ferguson of Leith, Scotland around 1880 and was taken over by the Queensland Naval Brigade in 1885, and served as a Gun Boat during the Russian scare.'
Oxley Parade: Surveyor-General John Oxley, returning from investigating Port Curtis in the ship 'Mermaid', called into Moreton Bay, where, in 1823, he recorded the discovery of a large river which he named Brisbane.
Pamphlet Street: Pamphlet was one of the three castaway cedar-getters, who, wrecked on Moreton Island,

crossed the South Passage in a canoe to Amity where he lived with the Aborigines for some months.

Parsons Street: Parsons was one of three castaway cedar-getters, who, wrecked on Moreton Island, crossed the South Passage in a canoe to Amity where he lived with the Aborigines for some months.

Petrie Street: Andrew Petrie (1798-1872), a pioneer who explored the Moreton region, contributed to the establishment of Brisbane Town. He was Engineer of Works and, in this capacity visited Dunwich frequently from 1837 to check the timber depot. His son, Thomas Petrie (1831-1910), knew the Aborigines and their customs well. Thomas Petrie's daughter, Constance, documented her father's stories in 'Tom Petrie's Reminiscences of Early Queensland', 1904.

Rainbow Crescent: 'His Excellency the Governor has been pleased to direct that the Island, forming the Southern Boundary of the Eastern Channel into Moreton Bay, shall be designated the Isle of Stradbroke in compliment to the Honble. Captain J. H. Rous, Commanding His Majesty's Ship "Rainbow", the first Ship of War which entered Moreton Bay. The point of Land on the Isle of Stradbroke, which is intended as the site of a Public Establishment opposite to Peel's Island, is named Dunwich, and the Anchorage where the "Rainbow" lay, Rainbow Reach.'

Rous Street: see Rainbow

Sally Street: see Bingle

Stradbroke Place: Captain J H Rous, Commander of His Majesty's Ship 'Rainbow', the first Ship of War to enter Moreton Bay, was the second son of the first Earl of Stradbroke (Viscount Dunwich).

Welsby Street: Thomas Welsby (1858-1941), an historian, Member of the Legislative Assembly of Queensland and Commodore of the Royal Queensland Yacht Club, recorded the life and history of early Moreton Bay.

APPENDIX III: Significant historical dates

1770 Lieutenant James Cook, sailing north along the eastern Australian coast in HMS 'Endeavour', named Point Lookout, Moreton Bay, and Cape Moreton.
1799 Lieutenant Matthew Flinders in HM sloop 'Norfolk' mapped northern Moreton Bay.
1803 Matthew Flinders, on his extraordinary open boat voyage in the cutter 'Hope', from Wreck Reef to Sydney to obtain help for the crews of the wrecked 'Porpoise' and 'Cato', put four men ashore in the vicinity of Cylinder Beach, Point Lookout, where they filled their water casks from a freshwater spring.
1822 Captain John Bingle, sent in the cutter 'Sally' to look for a large river supposed to exist between Port Macquarie and Sandy Cape, explored Moreton Bay but did not find a river.
1823 John Oxley recorded that Stradbroke was an island. He also rescued from Bribie Island the castaway timber-getters Pamphlett and Finnegan, who told him of a large river that flowed into Moreton Bay.
1824 Captain Penson, Surveyor-General John Oxley, and the botanist Allan Cunningham, in the brig 'Amity', rounded Cape Moreton to enter Moreton Bay, where they formed a settlement at Redcliffe. Captain Penson on his return to Sydney was the first to use the South Passage, the opening between Moreton Island and Stradbroke Island.
1825 'John Gray surveyed and buoyed the South Passage,...and from that time ships proceeding to the Brisbane River went past Amity Point and Dunwich, around the southern side of Peel Island, across to the "Green Islands"...to the river.'
Major Lockyer in the cutter 'Mermaid' made surveys in Moreton Bay.
1827 Governor Darling ordered the establishment of a depot at Dunwich (Green Point)
'...His Excellency the Governor has been pleased to direct that the Island, forming the Southern Boundary of the Eastern Channel into Moreton Bay, shall be designated the Isle of Stradbroke in compliment to the Honble. Captain J. H. Rous, Commanding His Majesty's Ship "Rainbow", the first Ship of War which entered Moreton Bay. The point of Land on the Isle of Stradbroke, which is intended as the site of a Public Establishment opposite to Peel's Island, is named Dunwich, and the Anchorage where the "Rainbow" lay, Rainbow Reach. The Channel between the Isle of

Stradbroke and Moreton Bay is named Rous Channel.'
1828 Captain Logan 'reported that the magazine at Dunwich was roofed and nearly completed, and the military and prisoners' barracks at Dunwich were finished and occupied.'
When the botanists Allan Cunningham and Charles Fraser visited Dunwich, they reported that the cotton growing there was neglected.
1831 The Dunwich depot was closed when it was found to be unsatisfactory.
1843 Bishop Folding organised the establishment, at Dunwich, of the first recorded Roman Catholic Mission to the Australian Aborigines.
1847 The wreck of the paddle steamer 'Sovereign' on the South Passage bar: 44 lives were lost.
A plan for the northern entrance to Moreton Bay was completed and sent to Sydney for approval.
1850 The ship 'Emigrant' with 64 cases of typhus aboard was quarantined at Dunwich.
Gustavus Birch resided at Amity 1850-1883; in his 1873 diary he recorded island life.
1853 A barque of 329 tons, the 'Countess of Derby', was wrecked on the South Passage Bar.
Ten Norfolk Island escaped convicts landed on Stradbroke. They terrorised the people of Moreton Bay until recaptured 16 days later.
1856 The 'Phoebe Dunbar' ran aground near Amity Point, but was refloated.
1858 Completion of the Cape Moreton lighthouse
1864 The Benevolent Asylum established in the old quarantine buildings at Dunwich
1874 Formation of the Moreton Bay Oyster Company
1886 The Surveyor-General's Office published a plan of 124 housing allotments for the Town of Amity Point.
The death of Toompani, who was buried in true tribal custom on the hill behind Amity Point
1892 When the Bribie Island Aboriginal Mission was closed, the people and their belongings were moved to Moongalba, near Dunwich.
1894 The 'Cambus Wallace', a barque of 1651 tons, was wrecked off Swan Bay.
1896 The sea broke through at Jumpinpin to create North and South Stradbroke Islands.
1902 The 133 ton brigantine 'Prosperity' of Sydney was wrecked half a mile southeast of Point Lookout.
Two men were lost.
1932 Construction of the Point Lookout lighthouse
1934 Bert Clayton commenced a regular beach bus service from Amity to Point Lookout.

1935 Hayles Bros of north Queensland built a kiosk at Amity Point, and commenced a launch service from North Quay, Brisbane. MV 'Mirimar' of 96 feet was the first launch on the run.
1936 Death of Sam Rollands, one of the last of the dugong fishermen
1942 An American liberty ship 'Rufus King' was wrecked on the South Passage Bar.
1943 The Australian hospital ship 'Centaur' was torpedoed by a Japanese submarine, 40 miles east of Brisbane. When a Dutch Dakota plane disintegrated and plunged into the sea between Shag Rock and Moreton Island, its six crew members perished. Myora Mission Reserve closed when the land was transferred to the Dunwich Benevolent Asylum.
1947 Removal of the Dunwich Benevolent Asylum to Sandgate. North Stradbroke Island was placed under the control of the Redland Shire.
1950 The first shipment of mineral sand from North Stradbroke
1956 The Moreton Bay Oyster Company was wound up.
1964 The vehicular ferry 'Myora' of Stradbroke Ferries Pty Ltd commenced a service between Redland Bay and Dunwich. V W Brook and A G Durbidge of the Hotel Stradbroke operated the first Stradbroke drive-on and drive-off vehicular ferry, which ran from Paxton Street, Cleveland.
1973 Stradbroke Ferries moved its operations to Toondah Harbour, Cleveland.
Riverside Ferry Co Pty Ltd, with the 'Riverside Venture', commenced a vehicular ferry service between Redland Bay and Dunwich.
1983 Commencement of Cleveland-Dunwich water taxi service
1988 Queensland government abandoned the Stradbroke bridge proposal.
2004 Centenary of Dunwich School

APPENDIX IV: Places of historical interest

Dunwich Cemetery

Dunwich was declared a quarantine station in 1850. In September of that year, the ship 'Emigrant' arrived with 64 cases of typhus. In the Dunwich Cemetery, 26 of the typhus sufferers are buried near the graves of Dr Ballow of Brisbane and the ship's Dr Mitchell who died attending to the typhus patients. When the quarantine facilities were moved to Peel Island, the government established the Dunwich Benevolent Asylum to cater for the old, homeless and ill. Many inmates lie in the cemetery, which contains around 9000 graves, many unmarked. It is possible for people tracing an ancestor that was an inmate in the Asylum to check the North Stradbroke Island Historical Museum's copy of the Asylum death register, 1886-1947.

Dunwich causeway

Under Captain Logan who established a stores depot at Dunwich, convicts constructed a stone causeway for the handling of goods off-loaded for transport to Brisbane on smaller craft. The northern wall which forms part of the present Dunwich jetty/ramp loading complex is still visible and is the oldest known convict marine structure in what is now known as Queensland.

Saint Mark's Church of England, Dunwich

Built in 1907, this small wooden structure was designed by Robin Dods, a renowned Brisbane architect. The church was built as a place of worship for the inmates of the Dunwich Benevolent Asylum.

Dunwich Public Hall

The hall was constructed in 1912 as the dining room for the Dunwich Asylum inmates. The interior walls are tiled to window height. As the food was brought by horse and dray from the kitchen some distance away, it was often cold when put in front of the inmates. During the war, the middle was sawn out of the hall's wooden floor, and the area was sandbagged to provide an air raid shelter. The building which now serves as a public hall is a good example of colonial architecture.

Old Institution buildings

The mineral sandmining company restored a section of the Dunwich Benevolent Asylum Ward 13. The building is used by Consolidated Rutile Ltd as a training centre. The front section of the North Stradbroke Island Historical Museum is built from recycled timber and windows from an old Dunwich Benevolent Asylum ward.

Ballow Park

In Ballow Park, Dunwich, the inmates played draughts on the big concrete draughts boards, shaded by huge Camphor Laurel trees.

Old Institution gas combustion engine

At the rear of the fenced yard of the Dunwich Pump Station at the Two Mile, Yerrol Creek, is the Institution's old gas combustion engine that drove three plunger water pumps that filled the water tank on the hill behind Dunwich. To preserve the old engine, the Redland Shire Council has housed the machine in an open shed.

Declared Aboriginal relic sites

In Queensland, all declared Aboriginal Relic Sites are protected under the 1967 Act. The Cultural Unity, Department of Natural Resources and Mines, controls the identification and protection of Aboriginal sites. Easily accessible sites can be seen at Polka Point, Dunwich, and on the turn-around near the Life Savers Club House, South Headland, Point Lookout.

Captain Cook seat

Cook, sailing up the east Australian coast, sighted and named Point Lookout in 1770 as a warning to mariners. Two hundred years later, interested citizens constructed the seat which is built of porphyry blocks from the dismantled fire brigade building in Ann Street, Brisbane. A time capsule was sealed into the construction.

Moongalba

Moongalba near Dunwich, the site of the old Aboriginal Mission (1892-1947), is now the burial site of the poet, Oogeroo Noonukul, or Kath Walker, and her son Vivian.

North's cattle dip

Mr Bill North ran cattle on the island to supply the Dunwich Institution with beef. He built a cattle dip, and dug a well at the Adder Rock corner on the high side of the present main road.

Myora Springs

This small pocket of island rainforest is where the castaway timber-getters, Parsons, Pamphlett and Finnegan, cut a tree to construct the dugout canoe in which they paddled to the mainland. It is also the site of the one-day battle between the soldiers and the Aborigines in 1833. Once, Myora was a popular watering spot for boating people, who, at high tide, rowed up the creek through the mangroves to where the freshwater gushed over a constructed wooden race into a small swimming hole.

Cylinder Beach

Cylinder Beach, Point Lookout, is where Matthew Flinders on his epic open boat trip from Wreck Reef to Sydney, in 1803, landed four men to replenish his water casks. Also in this sheltered bay, much of the material to construct the Point Lookout lighthouse was ferried ashore from the 'Cape York'. Back from the beach, a storage shed was built for the gas cylinders for the light.

'Prosperity' wreck

At times when the sea is calm, and when the sand scours away from the wrecked 133 ton brigantine 'Prosperity', which lies completely submerged half a mile southeast of South Head, Point Lookout, the ship is visible from the headlands.

'Rufus King' wreck

The rusting stern of the liberty ship 'Rufus King', wrecked in 1942 on the South Passage Bar, is no longer visible above the water line.

PEI Road

Under the government's Public Estate Improvement scheme, around about 1937 a road was built from Amity Point to Dunwich via Blue Lake. The PEI scheme was devised to employ men during a severe economic depression. The sealed section of Beehive Road from Amity to Dickson Way is part of the original PEI survey. The balance of the old road, now a 4 wheel drive track, offers fine island views. The track commences at the junction of the Amity-Beehive Road with Dickson Way.

Clayton's Old Road

Bert Clayton commenced a beach bus service from Amity Point to Point Lookout in 1934. When Hayles Cruises commenced a regular launch service from North Quay, Brisbane to Amity a year later, Clayton was faced with finding a route along which his service could operate to a timetable and not be dependent on tides. Along the razor-back ridge that runs parallel to the Amity-Point Lookout Swamp, he cut a narrow road. From this old road there are some good views of the swamp and Moreton Island. A lower road, skirting the swamp, was constructed later. This was named Clayton's Road.

Adam's Beach

This is an attractive beach at Dunwich. Its name derives from Benevolent Asylum days when male inmates used to go 'skinny dipping' there.

Frenchman's Beach

It is believed that the beach is named after an immigrant from French-speaking New Caledonia who landed on Frenchman's. From there he found his way to the Myora-Moongalba Mission.

Deadman's Beach

When during rough weather the brigantine 'Prosperity' was wrecked 800 metres south of Point Lookout in 1902, the cook and first mate were drowned. In 1956, a male skeleton was uncovered in a high dune at the southern end of the beach. The skeletal remains were reputed to be of one of the crew of the 'Prosperity'.

APPENDIX V: Scenic places

Eighteen Mile Swamp

This great freshwater swamp extends along Stradbroke's eastern road from Yarraman Valley to Swan Bay. It is a unique landform. Beach sand dunes fringe its seaward side and vast palm forests thrive in its southern region. This swamp is home to many species of water birds. Black Swans, Black Ducks, Eastern Swamphens, Coots, Pied Cormorants, and many types of herons and egrets are seen commonly. Musk Ducks, Jabirus, and White-eyed Ducks also occur in quiet refuges.

Amity and western swamps

Some of the Amity and western swamps which are fringed by mangroves are the habitat of interesting and specialised fauna and flora. The Swamp Orchid (*Phaius australis*), Water Mouse (*Xeromys myoides*), golden Swamp Wallaby (*Wallabia bicolor*), Mangrove Honeyeater (*Lichenostomus fasciogularis*), and the Striated Heron (*Butorides striatus*) are all found in these swamps, along with many more common species.

Blue Lake-Tortoise Lagoon National Park

The Aborigines named the lake Karboora, 'the silent pool'. A white sandy-bottomed, fast-flowing creek with steep fern-covered banks carries the Blue Lake overflow down to the Eighteen Mile Swamp.

Cairns Hill

This rocky outcrop, approximately 1 km south of Point Lookout, and accessible from the beach to walkers, offers excellent views of the Point Lookout headlands and the 35 km Main Beach. The original cairn was built by surveyors as a trig point. Visiting hikers have built further cairns, or added to existing structures.

Brown Lake

This lake is accessible to conventional drive vehicles. In spring, the yellow-flowering Pea (*Aotus* sp) and the purple-red Boronia (*Boronia falcifolia*) are spectacular. Brown Lake was known to the Aborigines as Bummel and Bummiera. There is a walking track from the lake to Myora.

Ibis Lagoon, Horseshoe Swamp, Tea-Tree Lagoon, Native Companion Lagoon, Duck Lagoon and South Lagoon

Located some distance from the townships, these are not easily accessible. Mineral sandmining operations have had a detrimental impact on some of these waterbodies.

Swan Bay and Never-Never (Duck) Creek

The Eighteen Mile Swamp drains into Swan Bay. The

shallow bay, creeks associated with it, saltmarsh flats, and mangrove-covered islands form the Swan Bay Sanctuary.

Point Lookout headlands and gorges

These superb rocky headlands and gorges (Mooloomba – 'rocks warmed by the sun', to the Aborigines) provide ideal vantage points for whale and sea bird watchers. On a clear day, Mt Warning is visible, and at night, the glow from the Gold Coast lights can be seen. When heavy swells force waves up a narrow rock crevice, Whale Rock blows. The Aborigines referred to the rock as Dunboonee.

Fern Gully

This deep-sided fern-covered gully, off Clayton's Road (the old Amity-Point Lookout sand track), is fed by a perched swamp at the head of the gully.

Swallow Lagoon

This is a small lagoon lying east of Dunwich in a deep depression formed by the surrounding hills and Mt Hardgrave. Energetic hikers find the trip worthwhile.

Lake Kounpee, Blaksley's Lagoon, Black Snake Lagoon, and Shag Lagoon

These lagoons lie south of Dunwich. Unfortunately, Kounpee, a perched lake, has been drained permanently by mineral sandmining fracturing the lake's impervious layer.

Wallum Creek and the Amity Swing Basin

Wallum Creek drains Amity's southeastern freshwater swamps into the Swing Basin, which was once an excellent anchorage for small boats. Silting has drastically reduced the size of the anchorage which is now only accessible at high tide. The extensive tidal flats of the basin provide yabbies for the angler.

Myora Springs

This is the only accessible patch of rainforest on the island. It is a refuge for the few rainforest species of animals on the island. The mangroves which fringe the rainforest at Myora support a population of one of the rarest and least-known mammals in Australia, the Water Mouse (*Xeromys myoides*). See Appendix IV for information on the historical significance of Myora Springs.

Welsby's Lagoons

These lagoons lie approximately 10 km from Point Lookout, just to the western side of Dickson Way which links Point Lookout and Dunwich. They are close to the road and are accessible on foot.

Mt Scott, Mt Willes and South Hill

From these sandy steep hills it is possible to view much of Moreton Bay. These ancient dunes have been subjected to mineral sandmining. The Stradbroke Energex power line from Beenleigh, crossing Russell Island, spans Canaipa Passage to a pylon on Mt Willes.

APPENDIX VI:
Early maps

(from Covacevich, Durbidge and McKinnes, in *Focus on Stradbroke*, 1984)

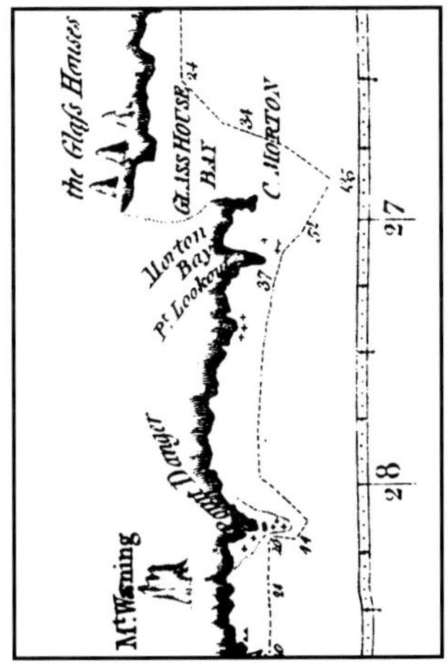

Plate 138: James Cook

Plate 138: The coastline from Mt Warning to Wide Bay after James Cook, 1770 (Evans 1969). Separations from the present Southport Spit and Moreton Island, which presumably existed, were not observed, probably because the 'Endeavour' was too distant from the shore. On May 17, 1770 he recorded: 'the northermost land bore N. by W., the breakers, N.W. by W. distant four miles, and the northermost land set at noon, which formed a point, and to which I gave the name of Point Look-out, W. distant five or six miles, in the latitude of 26°6'. On the north side of this Point, the shore forms a wide open bay, which I called Moreton's Bay...'

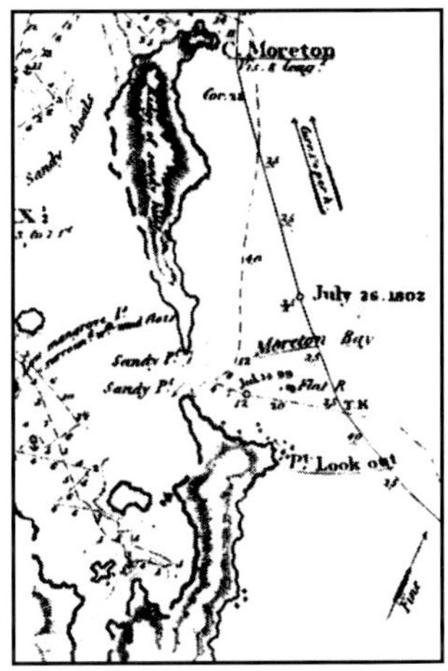

Plate 139: Matthew Flinders

Plate 139: Matthew Flinders, in 1794 (Steele 1983) added an 'e' to Moreton, and refined knowledge of the shape of the northern part of Stradbroke as well as the Moreton Bay area generally. His chart of the area was published in 1814 (Steele 1983). The break between the present Southport Spit and the southern end of South Stradbroke was not recorded at that time.

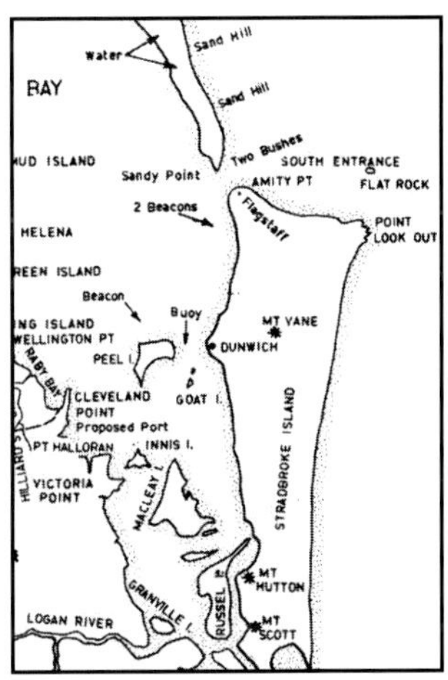

Plate 140: Robert Dixon

Plate 140: Robert Dixon's 1842 map of the Moreton Bay District 'compiled from authentic surveys and containing all the latest discoveries made by exploring parties...most respectfully dedicated to his Grace the Duke of Cleveland.' Redrawn by Steele (1975) and reproduced with his permission.

Prior to the release of land for sale to free settlers in the Moreton Bay District, three assistant surveyors, Robert Dixon, James Warner and Granville C Stapylton, were sent from Sydney in 1839. Assisted by convicts, the surveyors split into three parties. In 1840, Stapylton and his chainman, Tuck, were murdered by Aborigines in the Mt Lindsay area. What remained of Stapylton's beheaded body was transported to Brisbane for burial in the vicinity of Roma and Quay Streets. After a disagreement with Lieutenant Gorman, Dixon was dismissed in 1841 from his position as Assistant Surveyor (Kitson, personal communication; Steele 1975).

APPENDIX VII:
Rainfall figures for Point Lookout, 1947-1980 (mm)

	Jan.	Feb.	Mar.	Apr.	May	Jun.	Jul.	Aug.	Sep.	Oct.	Nov.	Dec.	Total
1947	-	198	271	150	195	12	10	56	94	65	102	158	1310
1948	147	38	162	174	184	254	34	167	123	-	72	51	1404
1949	77	300	192	178	191	169	98	96	128	216	92	69	1807
1950	133	401	121	255	103	176	244	263	28	126	70	79	1338
1951	550	100	302	60	93	183	5	45	35	56	6	59	1495
1952	44	138	305	134	85	145	147	125	69	170	31	55	1448
1953	172	274	259	29	51	5	38	173	29	126	54	79	1289
1954	192	761	166	118	269	55	174	237	156	181	53	111	2472
1955	75	41	312	94	234	77	120	11	67	72	5	331	1437
1956	56	670	276	147	234	83	71	6	69	54	24	278	1979
1957	229	115	269	63	15	95	205	152	18	79	67	25	1332
1958	89	163	248	313	68	356	1	168	132	53	66	195	1852
1959	168	80	307	166	206	129	254	39	131	111	178	109	1876
1960	219	168	103	44	62	77	221	18	1	104	96	68	1180
1961	157	127	155	119	120	44	62	114	101	167	292	-	1458
1962	354	103	141	243	83	30	399	133	55	38	167	171	1918
1963	162	57	531	195	338	163	18	201	20	149	149	380	2362
1964	154	390	309	178	159	50	36	34	80	104	119	99	1711
1965	75	41	25	154	101	178	194	100	46	52	7	268	1241
1966	19	111	82	218	33	351	41	170	84	100	119	111	1438
1967	313	124	376	108	221	598	87	110	3	143	54	130	2266
1968	293	51	54	60	110	73	67	96	14	5	42	73	938+
1969	16	72	53	61	249	27	162	48	12	496	186	120	1502
1970	151	169	163	37	66	32	11	28	70	36	119	339	1221
1971	319	361	97	66	191	183	71	83	47	54	100	146	1715
1972	224	350	373	183	252	162	5	73	34	470	106	62	2293
1973	61	321	63	137	181	53	514	47	64	245	80	105	1870
1974	759	106	441	400	217	427	15	75	67	71	331	27	2936*
1975	238	201	110	121	164	117	47	150	109	146	120	143	1666
1976	99	276	539	191	326	214	91	13	75	113	246	50	2232
1977	211	120	146	148	126	125	81	31	28	84	184	44	1329
1978	180	132	305	107	172	27	73	79	59	56	165	171	1528
1979	99	68	81	199	53	113	197	2	3	82	99	46	1041
1980	119	149	7	43	582	80	61	96	-	147	68	194	1512

Rainfall (mm): (1997) 1263; (1998) 1216; (1999) 2517; (2000) 1063; (2001) 1059; (2002) 1187; (2003) 1527

Plate 141: In 1997 the Bureau of Meteorology established a recording station on the Point Lookout headland. Temperature, wind, cloud, rainfall details and state of the sea are recorded every three hours from 6am to 6pm. Residents are responsible for recordings and for transferring the information to the Bureau. Christine Durbidge is a local weather recorder.

REFERENCES

1 AMITY PROGRESS ASSOCIATION. 1978. 'A pilot survey of attitudes on North Stradbroke Island to the proposal to construct a bridge to link the island and the mainland'. (Unpublished report)
2 ARCHER, M. 1972. Domestic cats and dogs – a danger to the Australian fauna. West Aust Nat 12 (4): 85-87
3 BACKHOUSE, A. 1843. 'A Narrative of a Visit to the Australian Colonies'. (Hamilton Adams: London)
4 BAILEY, F M. 1909. 'Comprehensive catalogue of Queensland Plants'. (Government Printer: Brisbane)
5 BARRY, D H and CAMPBELL, P R. 1977. A survey of the mammals and herptiles of Fraser Island, with comments on the Cooloola Peninsula, North Stradbroke, Moreton, and Bribie Islands. Occas Papers in Anthropology 8: 147-177
6 BARRY, D H and VERNON, D P. 1977. The birds of Fraser Island, with comments on the Cooloola Peninsula, Stradbroke, Moreton and Bribie Islands. Occas Papers in Anthropology 8: 179-205.
7 BEACH PROTECTION AUTHORITY, 1979. 'Annual Report'
8 BECKMAN, G G. 1975. Southern Moreton Bay – past, present and future. Proc Roy Soc Qd 86: 133-139
9 BEAGLEHOLE, J C (ed). 1955. 'The Journals of Captain James Cook Vol. 1 – The Voyage of the Endeavour 1768-1771'. (Cambridge University Press: Cambridge)
10 BENSINK, A H A and BURTON, H. 1975. North Stradbroke Island – a place for freshwater Invertebrates. Proc Roy Soc Qd 86: 29-45
11 BENUSSI, G. 1975. Genesis of the island. Proc Roy Soc Qd 86: 3-8
12 BRYDEN, M M. 1978. Presidential address: Whales and whaling in Queensland waters. Proc Roy Soc Qd 89: v-xvii
13 BUNT, J S. 1981. Mangrove forests – a national asset to be guarded. Aust Fisheries 40 (I): 14-17
14 CHALLONER, M C. 1975. The local authority role in the future of Stradbroke Island. Proc Roy Soc Qd 86: 121-123
15 CHITTEEBOROUGH, R G. 1981. Habitats in coastal waters need protection. Aust Fisheries 40 (I): 10-13
16 CILENTO, R and LACK, C. 1959. 'Wild white men in Queensland'. J Roy Hist Soc Qd 6 (I): 73-93. (Special journal to mark the centenary of Queensland 1859-1959)
17 CLARIDGE, G and DAVIES, K (eds). 1977. 'The Future of Moreton Island'. (Queensland Conservation Council: Brisbane)
18 CLARIDGE, G. 1978. 'The Sandmining Handbook' 2nd ed (Queensland Conservation Council: Brisbane)
19 CLARK, S S. 1975. The effect of sandmining on coastal heath vegetation in New South Wales in 'Managing Terrestrial Ecosystems', (eds) Kikkawa, J and Nix, H A. Proc Ecol Soc Aust 9: 335 pp
20 CLIFFORD, H T and SPECHT, R L. 1979. 'The Vegetation of North Stradbroke Island' (University of Queensland Press: Brisbane)
21 COLLIVER, F S and WOOLSTON, F P. 1975. The Aborigines of Stradbroke Island. Proc Roy Soc Qd 86: 91-104
22 COLLIVER, F S and WOOLSTON, F P. 1978. Brisbane's early Aboriginals in 'Brisbane Retrospect, eight Aspects of Brisbane History'. (Library Board of Queensland: Brisbane)
23 COOK, CAPT J, see 'The Journals...' (ed) Beaglehole, J C
24 'COURIER-MAIL', 1980. editorial, May 12
24a 'COURIER-MAIL', 1981. article p 21
25 COVACEVICH, J. 1980 (2nd ed). 'Snakes of Brisbane' (Queensland Museum: Brisbane)
26 COVACEVICH, J and ARCHER, M. 1975. The distribution of the Cane Toad, Bufo marinus, in Australia and its effects on indigenous vertebrates. Mem Qd Mus. 17 (2): 305-310
27 COVACEVICH, J. and INGRAM, G J. 1975. Reptiles of Stradbroke Island. Proc Roy Soc Qd 86: 55-60
27a COVACEVICH, J, OGILVIE, A and WALKER, F. 1996. 'Stone walls do not a prison make...' profiles of prisoners on Stradbroke Is, Qld, in Cocacevich et al (eds) History, Heritage and Health. Proceedings of the 4th Biennale Conference of the Aust Soc of the Heritage of Medecine
28 CRIBB, A B and CRIBB, J W. 1974. 'Wild Food in Australia'. (Fontana, Collins: Melbourne)
29 CRIBB, A B and CRIBB, J W. 1981. 'Useful Wild Plants in Australia'. (Collins: Sydney)
30 DAHMS, E G. 1978. A checklist of the types of Australian Hymenoptera described by Alexandre Arsene Girault: 1. Introduction, acknowledgments, biography, bibliography and localities. Mem Qd Mus 19 (2): 127-190
31 DAVIDSON, P M. 1976. Birds in pine forests. A brief study of the relationships between bird populations and pine plantation habitats at Kowen Forest, A.C.T. Studies in Forest Environment no. 1. (Aust Government Publishing Service: Canberra)
32 DICKSON, B. 1979. The Spanish galleon, or was it? (Unpublished document circulated by the author)
32a DICKSON, B. 1977. Address on the Dunwich Benevolent Institution to the Queensland Women's Historical Association. (Unpublished document circulated privately by the author)
33 BOWLING, R M. 1979. The mangrove communities of Moreton Bay in 'Northern Moreton Bay Symposium', eds Bailey, A and Stevens, N C. (Royal Society of Queensland: Brisbane)
34 DURBIDGE, E. 1975. Introduction to the island. Proc Roy Soc Qd 86: 1-2
35 DWYER, P D, KIKKAWA, J, and INGRAM, G J. 1979. Habitat relations of vertebrates in subtropical heathlands of coastal southeastern Queensland, in 'Ecosystems of the World, 9A, Heathlands and related Shrublands', ed. Specht, R L (Elsevier Scientific Publishing Company: Amsterdam)
36 ERICKSON, R. 1978. 'Plants of Prey in Australia', (University of Western Australia Press: Nedlands)
37 EVERIST, S L. 1974. 'Poisonous Plants of Australia'. (Angus and Robertson: Sydney)
38 EVERIST, S. 1975. Stradbroke Island as a place for plants. Proc Roy Soc Qd 86: 21-23
39 FITZGIBBONS, A. 1975. The economics of sandmining and other island industries. Proc Roy Soc Qd 86: 105-109
40 FLINDERS, M, 1803. Log of the 'Investigator',

'Hope', and 'Cumberland' (Part 3 of Flinders' Logs), from the Mitchell Library, Sydney
41 FLINDERS, M, 1814. 'A Voyage to Terra Australis'. Vols 1, 2. (G W Nichol: London)
42 GILL, A M, 1975. Australia's mangrove enclaves: A coastal resource, in 'Managing Aquatic Ecosystems', eds Nix, H A and Elliot, M A. Proc Ecol Soc Aust 8: 173 pp
43 GUTTERIDGE, HASKINS AND DAVEY. 1975. 'Coastal Management. Queensland-New South Wales Border to Northern Boundary of Noosa Shire'. Vols 1-4. (Co-ordinator-General's Department: Brisbane)
44 HARDLEY, R. 1975. The social life of Stradbroke Island Aborigines. Proc Roy Soc Qd 86: 141-146
44a HALL, H J. 1984. Fishing with Dolphins? Affirming a traditional Aboriginal fishing story in Moreton Bay. SE Qld (Boolarong Publications, Brisbane and Stradbroke Island Management Organisation, Amity Point)
45 HIGHWAY PLANNING BRANCH, 1978. 'Proposed road link to North Stradbroke Island Environmental Study Advice'. (Main Roads Department: Brisbane)
46 HIGHWAY PLANNING BRANCH, 1978. 'Redland Shire Proposed Road Link to North Stradbroke Island'. Report prepared for discussion at public meeting 9.3.1978. (Main Roads Department: Queensland)
47 HIGHWAY PLANNING BRANCH, 1979. 'Redland Shire Proposed Roadlink to North Stradbroke Island'. Impact Assessment Study Report Stage I. (Main Roads Department: Brisbane)
48 HINZE, R J. 1978. Proposed road link between the mainland and North Stradbroke Island. The Helmsman Feb. 1978:19-22
49 HINZE, R J. July, 1979. (Letter to the Stradbroke Island Management Organisation)
50 HODGES, J C et al. 1977. 'Off-road Vehicles. Impact on the Australian Environment'. Third Report of the House of Representatives Standing Committee on Environment and Conservation. (Australian Government Publishing Service: Canberra)
51 HODGES, J C et al. 1980. 'Management of the Australian Coastal Zone. Report from the House of Representatives Standing Committee on Environment and Conservation'. (Australian Government Publishing Service: Canberra)
52 HOPE, R M, BRINE, J M, DUNPHY, M K, VALLANCE, N K, WALKER, R N, WEBB, L J, WRIGHT-McKINNEY, J and YENCKEN, D G D. 1974. 'Report of the National Estate. Report of the Committee of Enquiry into the National Estate'. (Australian Government Publishing Service: Canberra)
53 HUXLEY, W G S, HARROLD, A G and WYVILL, L F. 1980. The application of environmental legislation in Queensland. (Cooloola Committee: Box 238, North Quay, Queensland)
54 INGRAM, G J and CORBEN, C J. 1975. The frog fauna of North Stradbroke Island with comments on the 'acid' frogs of the wallum. Proc Roy Soc Qd 86: 49-54
55 INTERDEPARTMENTAL COMMITTEE TO ADVISE ON THE PREPARATION OF A MANAGEMENT PLAN FOR FRASER ISLAND, 1978. 'Fraser Island Management Plan'. (Publisher not identified on report)
56 KIKKAWA, J, 1975. A place for land vertebrates. Proc Roy Soc Qd 86: 47-48
57 KIKKAWA, J, INGRAM, G J, and DWYER, P D. 1979. The vertebrate fauna of Australian heathlands – an evolutionary perspective, in 'Ecosystems of the World, 9A, Heathlands and related Shrublands', ed Specht, R E (Elsevier Scientific Publishing Company: Amsterdam)
58 LANG, J D. 'Cooksland in north-eastern Australia. The Future Cotton Field of Great Britain' (Johnson reprint, 1970)
59 LAVERY, H J. 1979. Presidential Address: Conservation and change. Proc Roy Soc Qd 90: 1-12
60 LAYCOCK, J W. 1975. Hydrogeology of North Stradbroke Island. Proc Roy Soc Qd 86: 15-19
61 LEAR, R and TURNER, T. 1977. 'Mangroves of Australia'. (University of Queensland Press: Brisbane)
62 LEE-MANWAR, G, ARTHINGTON, A H A, and TIMMS, B V. 1980. Comparative studies of Brown Lake, Tortoise Lagoon, and Blue Lake, North Stradbroke Island. I Morphometry and origin of the lakes. Proc Roy Soc Qd 91: 53-60
62a LEWIS, E, 2003. Water Mining on North Stradbroke Island – Development, Management and the Role of the Public. (Unpublished document)
63 LOCKYER, N. 1920. Exploration by Major Edmund Lockyer of the Brisbane River in 1825. Hist Soc Qd J. 2: 54-73
64 MACNAE, V V. 1968. in Lear, R and Turner, T. 1977
64a MACKANESS, G. 1956. The discovery and exploration of Moreton Bay and the Brisbane River, Parts I, II (Printed privately by the author: Brisbane)
65 MARKS, E N. 1973. Saltmarsh mosquito control. Operculum 3(5-6): 87-88
66 MARTIN, J H D. 1975. A list of mammals from Stradbroke Island. Proc Roy Soc Qd 86: 73-76
67 MATHER, P, ed. 1976. 'The National Estate, Moreton and Wide Bay – Burnet Regions. A Report for the Co-ordinator-General'. (Queensland Museum: Brisbane)
68 McARTHUR, K. 1978. 'Pumicestone Passage, a living Waterway'. (Kathleen McArthur: PO Box 71, Caloundra)
69 McEVOY, J S. 1975. A plan for the preservation of habitat for the mammalian and avian fauna of North Stradbroke Island. Proc Roy Soc Qd 86: 77-79
70 McEVOY, J S. 1979. The taxonomic status of the Red Stradbroke Wallaby. Mammalia 43(4): 582-583
71 McHARG, I E. 1971. 'Design with Nature'. (Doubleday and Co, for the American Museum of Natural History: New York)
72 McINTYRE, K G. 1977. 'The Secret Discovery of Australia; Portuguese Ventures 200 years before Captain Cook' (Souvenir Press: Great Russell Street, London)
73 MORETON REGION GROWTH STRATEGY INVESTIGATION, 1974. (Queensland Government: Brisbane)
74 MOSLEY, G (ed) 1978. 'Australia's Wilderness: Conservation Progress and Plans'. (Australian Conservation Foundation: Melbourne)
75 O'KEEFE, M. 1975. Some aspects of the history of Stradbroke Island. Proc Roy Soc Qd 86: 85-89
77 OWEN, P C. 1981. Shoreline management and the

fisherman – trying to avoid competition and conflict. Aust Fisheries 40(1): 18-21
78 PALMER, V. 1944. 'The Passage'. (Robertson and Mullins: Melbourne)
79 PATTERSON, R. 1980. Encouraging sightings of humpback whales off cast coast. Aust Fisheries, 39(4): 8-10
80 PETRIE, C C. 1904. 'Tom Petrie's Reminiscences of early Queensland (dating from 1837) recorded by his Daughter'. (Watson and Ferguson: Brisbane)
81 PIXLEY, A J. 1970. Shipwrecks in Queensland and adjacent waters. J Roy Hist Soc Qd 9(1): 151-162
82 PONOSOV, V V. 1964. 'Results of an archaeological survey of the southern region of Moreton Bay and of Moreton Island 1963-1964'. (University of Queensland, Department of Psychology: Brisbane)
83 REDLAND SHIRE COUNCIL STRATEGIC PLAN FOR NORTH STRADBROKE ISLAND, in Challoner, M C. 1975
83a RAHNSLEBEN scrapbook, Fryer Library, University of Queensland, St Lucia, Brisbane
84 REID, F. 1954. 'The Romance of the Great Barrier Reef. (Angus and Robertson: Sydney)
85 REYE, E J. 1973. Midges and mangroves. Operculum 3(4): 31-34
86 ROBERTS, G. 1979. 'The Birds of southeastern Queensland'. (Queensland Conservation Council: Brisbane)
88 ROBINS, R P. 1979. Moreton Island Aborigines, a preliminary consideration and bibliography, in 'Northern Moreton Bay Symposium'. eds Bailey, A and Stevens, N C. (Royal Society of Queensland: Brisbane)
89 ROGERS, R W. 1975. North Stradbroke Island: Resources and Management in 'Managing Terrestrial Ecosystems' eds Kikkawa, J and Nix, H A. Proc Ecol Soc Aust 9: 335pp
90 ROUTLEY, R and ROUTLEY, V. 1973. 'The Fights for the Forests, the Takeover of Australian Forests for Pines, Woodchips, and intensive Forestry'. (Australian National University Press: Canberra)
91 SAUNDERS, G W. 1975. Presidential Address: Variety and diversity. Proc Roy Soc Qd 86: v-xiv
92 SAUNDERS, G W. 1980. National Parks and Wildlife Service poster
93 SHARP, A. 1963. 'The Discovery of Australia'. (Oxford University Press: Amen House, London)
94 SLACK, A. 1979. 'Carnivorous Plants'. (A H and A W Reed: Sydney)
95 SMITH, P. 1977. A value analysis of wilderness. Search 8 (9)
96 SOIL CONSERVATION BRANCH OFFICERS, 1974. 'Moreton Region: Non-urban Land Suitability Study'. Part I. (Technical Bulletin 11, Division of Land Utilization, Department of Primary Industries, Brisbane)
97 SPFCHT, R L. 1975. A place for teaching biology. Proc Roy Soc Qd 86: 81-83
98 SPECHT, R L. 1975. The effect of fertilizers on sclerophyll (heath) vegetation – the problems of revegetation after sand-mining of high dunes. Search 6(11-12): 459-461
99 SPECHT, R L, ROE, E Mand BOUGHTON, V H (eds). 1974. Conservation of major plant communities in Australia and Papua New Guinea. Aust J Botany, Supplementary Series No 7
100 STEELE, J G. 1975. "Brisbane Town in Convict Days 1824-1842'. (University of Queensland Press: Brisbane)
101 STOCK, E J. 1978. A bridge would 'kill' Stradbroke. 'Courier-Mail', March 7: 5
102 STROUD, R H. 1973. Environmental, recreational, and commercial relationships within fisheries management. Operculum 3(5-6): 84-86
103 THATCHER, A C and WESTMAN, W E. 1975. Succession following mining on high dunes of coastal south-east Queensland, in 'Managing Terrestrial Ecosystems'. eds Kikkawa, J and Nix, H A. Proc Ecol Soc Aust 9: 335pp
104 THORPE, REV O. 1950. 'The first Catholic Mission to the Australian Aborigines'. (Pellegrini and Co: Brisbane)
105 THOMSON, A K (ed) 1967. 'The Collected Works of Thomas Welsby'. Vols. 1 and 2 (The Jacaranda Press: Brisbane)
106 THOMPSON, C H. 1975. Coastal Areas of southern Queensland – some land use conflicts. Proc Roy Soc Qd 86: 109-120
107 THOMSON, J M. 1975. A place for fishes. Proc Roy Soc Qd 86: 25-27
108 THOMPSON, C H and WARD, W T. 1975. Soil landscapes of North Stradbroke Island. Proc Roy Soc Qd 86: 9-14
109 VAN DYCK, S, BAKER, W B, and GILLETTE, D D. 1979. The False Water Rat Xeromys myoides on Stradbroke Island, a new locality in southeastern Queensland. Proc Roy Soc Qd 90: 84
110 VERNON, D P and MARTIN, J H D. 1975. Birds of Stradbroke Island. Proc Roy Soc Qd 86: 61-72
111 WALKER, K. 1972. 'Stradbroke Dreamtime'. (Angus and Robertson: Sydney)
112 WATKINS, G. 1891. Notes on the Aborginals of Stradbroke and Moreton Islands. Proc Roy Soc Qd 1: 110-129
113 WEBB, L J. 1980. Royal Society of Queensland after-dinner peroration, in 'Contemporary Cape York Peninsula'. (eds) Stevens, N C. and Bailey, A (Royal Society of Queensland: Brisbane)
114 WEBB, L J, et al. 1970. Assessment of sand stabilization procedures on North Stradbroke Island. (Unpublished report)
115 WELSBY, T. see '...Collected Works...'. ed Thomson, A K
116 WHITEHOUSE, F W. 1971. (Unpublished letter)
117 'THE ROAD AHEAD'. April-May, 1981, p16
118 ANONYMOUS. 1981. Fish species from inshore areas vital for Queensland fisherman. Aust Fisheries 40 (I): 41-43
119 ATS CONSULTANTS. March 2003. Redland Shire Sustainable Tourism Strategy Background and Discussion Papers

INDEX

Bat Cave, Point Lookout (The late Bob Pattison)